Computation for Process Engineers

Computation for Process Engineers

G. L. WELLS and P. M. ROBSON
University of Sheffield

A Halsted Press Book

John Wiley & Sons
New York

First published by
Leonard Hill Books
a division of
International Textbook Company Limited
24 Market Square, Aylesbury, Bucks. HP20 1TL

Published in the USA 1973
by Halsted Press, a Division of
John Wiley & Sons, Inc., New York

Library of Congress Cataloging in Publication Data

Wells, G L

Computation for process engineers.

"A Halsted Press Book."

Includes bibliographical references.

1. Electronic data processing—Chemical processes.
I. Robson, Peter Murray, joint author. II. Title.
TP155.7.W43 660.2'81'02854 73–6089
ISBN 0-470-93319-4

Printed in Great Britain by
Billing & Sons Limited, Guildford and London

Contents

List of Figures

Preface

In any book on computation which relates to a large field such as Process Engineering there are bound to be omissions and when the aim is one of introduction to the computational aspects only, the problem of which areas to select for illustration is even more difficult. As in the book *Process Engineering with Economic Objective*, I indicated many examples without showing or discussing the computational aspects of the problem, the present book may be considered as a supplement to that text.

It is assumed that the reader has a knowledge of Fortran programming and the numerical methods indicated are only those relevant to the examples in the book. The aspects upon which we have concentrated are the jargon used in computation as relevant to process engineering, the philosophy of such use and the assembly of the lengthy programs which are a feature of process engineering applications. Of course the latter do appear somewhat fearsome as flowcharts but by keeping to programs generally available it is hoped that the reader wishing to gain further expertise will attempt to use such procedures on the machine, thereby learning for himself that in practice the problem is by no means as difficult as might seem solely by reading.

I would like to acknowledge the assistance given in certain areas of this work by my research students, Messrs Burgess, Dawson and Makin, undergraduates Buck, Chester and Lim, colleagues at Sheffield University, Drs Burley, Swithenbank, Lafferty and Radley, initial assistance given with the program GEMCS by Drs Flowers and Whitehead of Leeds University, and initial instruction on simulation by Professor Johnson of Western Ontario University.

G. L. W.
August 1973

Structure of the Book

The basic approach is to follow the cyclic pattern of design as indicated below.

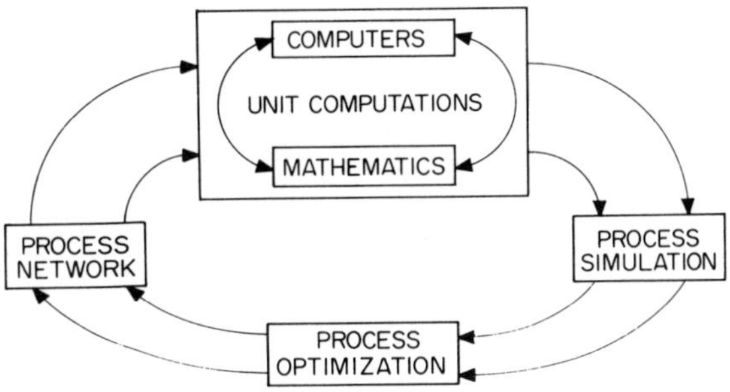

Starting with a brief description of some computing terms, chapter 1, we introduce unit computations by categorizing them in terms of strategy of use, chapter 2. This in turn is cyclic in nature following project development and usually follows the pattern noted above. Mathematics is used in conjunction with the computer to model the unit as efficiently as necessitated by the use, and some numerical and optimization methods are given in chapter 3. This list is by no means complete, in particular largely omitting differential equations. Otherwise the book would quickly be dominated by the problems of mathematical modelling. In chapter 4 the opportunity is taken to discuss in greater detail a specific unit computation, distillation, before considering the vital aspect, the flow of information, chapter 5. The sections are integrated together by process simulation, chapter 6; and some aspects of optimization of larger systems is discussed in chapter 7. The book concludes by discussing aspects of the use of computers for process engineering in the chemical industry. In content this last chapter contains items worth noting both at the beginning and end of the book and it is suggested to the reader that he might briefly skip through its contents at the beginning of his studies.

1 Computing Terms

1.1 Introduction

Many of the problems encountered in process engineering, whether they are concerned with design or analysis of results, require that numerical calculation be performed for their solution. This book outlines how some problems are being tackled utilizing the calculating power of digital computers. It is assumed that the reader has elementary programming knowledge of the Fortran language. As the book is an introduction to the subject an attempt is made to keep the chemical engineering and numerical methods content at a level which might be expected of a student commencing to apply his knowledge of computing.

Probably the most fundamental factor to be appreciated when concerned with computers is the limitations of the machine. The popular concept of a computer being an 'electronic brain' has led to the opinion that it possesses abilities that no machine could ever achieve. Computers can in no way provide a substitute for human intelligence and decision; they are only a tool which the engineer or technologist has available to help him tackle problems. Nevertheless, their range of applications is already vast and will no doubt increase ever more rapidly.

Before any attempt can be made to produce a computer program for solution of a particular problem the user must define the nature of the problem and what is required in the form of the solution. This might appear to be obvious and trivial but the importance of problem definition and goal identification cannot be over-emphasized when a computer solution is required. Both conditions imply a complete understanding of all aspects of the problem and also the conditions under which a solution is to be attained. Then, having become fully acquainted with the problem, it is necessary to outline the approach to be employed in the solution. This involves a mathematical knowledge of the relevant methods of analysis and a logical ordering of the steps necessary for the solution.

1.2 Flowcharts

Flowcharts are a means of illustration of the order of operations to be executed. Elementary programming theory will have introduced most readers to flowcharts but the temptation to avoid using them for small well-defined problems often extends to the larger more complex problems, and it must be stressed that their usage will greatly reduce the time spent on developing the program as well as providing a clear record of how the program has been produced. Also, use of a flowchart can be invaluable for checking the program logic when attempting to locate errors.

Some of the symbols adopted in flowchart construction are as follows:

An oval for the position at which a program starts or finishes.

A rectangle for statements involving mathematical calculations or equality relationships.

A diamond for the position at which a quantitative decision must be made. The lines leaving the box are labelled with the decision results that cause each path to be followed.

A trapezoid for read input and write output statements.

A circle for connection points in a flowchart where a connecting line would be untidy.

A simple illustration of flowcharting involves the evaluation of a step function,

$$y = \begin{cases} 14.23 & \text{if} \quad 0.0 \leqslant x < 6.0 \\ 19.18 & \text{if} \quad 6.0 \leqslant x < 15.0 \\ 27.06 & \text{if} \quad 15.0 \leqslant x < 32.0 \end{cases}$$

where with a value of x supplied to the program it is necessary to find the corresponding value of y. The flowchart is shown in figure 1.1.

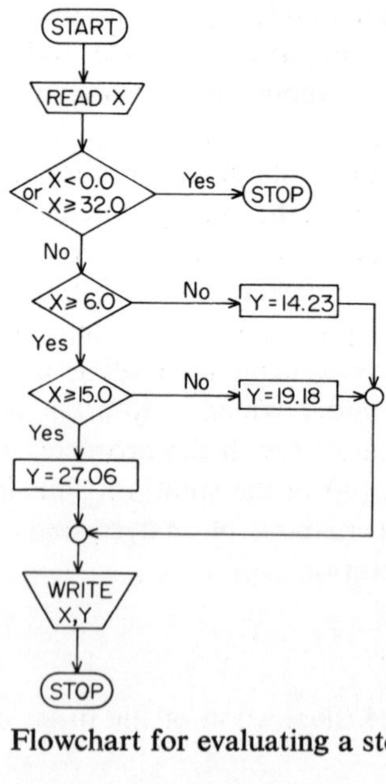

Figure 1.1 Flowchart for evaluating a step function.

1.3 Definition of an algorithm

An ordered set of steps in sequence as illustrated by a flowchart, can be regarded as the structure of an algorithm. A formal definition of an algorithm is as follows. A procedure for solving a problem, **P**, on a device, **D**, is called an algorithm for **P** on **D** if, after the performance of a finite number of steps, either a solution to **P** is obtained or **D** determines and reports that **P** has no solution (Sharpe, 1972).

The description of an algorithm requires that there is a language available which is acceptable to the device which will execute it, and that a number of precise steps, in terms of instructions which the device can perform, are ordered in a sequence, so that the performance of these instructions leads to a solution after a finite number of steps. Nowadays a great deal of research work concludes with the production of an algorithm, which if it can be accepted with confidence, means that the problem in hand needs no further investigation since a method of solution has become available.

1.4 Organization and operation of computer hardware

The term computer hardware is the name given to all the physical pieces of apparatus that go to make the computer system. Software is used to describe the operating systems, programs, etc, that are fed into or stored in the machine in whatever form.

It is of some value to understand what the automatic computer should and can do. The manipulation and representation of numbers is the most obvious requirement, but added to these the computer must be capable of recording and storing values, identifying and acting according to instructions as well as selecting the next instruction in order to proceed. With digital computers the representation of numbers by the system of radix two is known as binary notation. This implies that compared with the conventional denary system the actual number of digits needed to represent a number is large, but even so there are a number of features which make the binary system very suitable for computer use. The most important is that a binary digit can be represented by a 'two-state' device. However, this implies that all elementary operations within the computer must be dependent upon a choice of only two values, all information being represented in the computer by the states of individual circuit elements, i.e. each element has only two states equivalent to 0 or 1, implying in practice that either it is or is not conducting electricity. This introduces the concept of Boolean notation which refers to two valued quantities, such as the binary number digits by a symbol and its complement as a means of expressing more simply the logical function to be performed.

An appreciation of how a digital computer works can be obtained with the help of a model of the computer shown in figure 1.2.

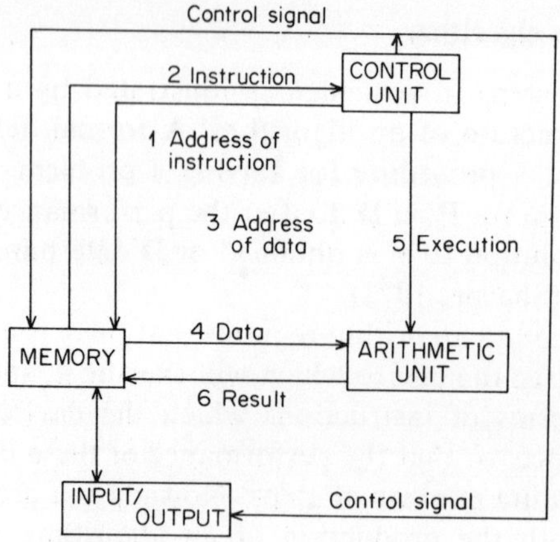

Figure 1.2 Model of the computer.

The sequence of operations is:

(1) Control unit specifies address of instruction to be brought from memory unit.

(2) Instruction stored at that address is sent to control unit and decoded.

(3) Control unit specifies address of required data on the basis of the decoded instruction.

(4) Data is sent from specified address in memory to arithmetic unit.

(5) Control unit instructs arithmetic unit to perform the operation decoded in step 2.

(6) Result of the calculation is stored in the memory location specified in the subsequent instruction.

Thus the computer must possess input/output units to allow the user to enter or extract information, an arithmetic unit to perform calculations, a memory unit to store instructions, data and intermediate results, and a control unit to direct overall action.

Before a computer can be constructed to perform arithmetical operations, building blocks are needed to carry out the operations as represented by the appropriate Boolean expressions. Such building blocks are known as 'gates' and the ones most often used nowadays to construct the logic circuits are NAND and NOR gates. These two gates are developments from the simple AND and OR functions and differ only in the fact that their respective outputs have been inverted, e.g. where an AND gate gives an output of 1 a NAND gate gives an output of zero. Table 1.1 shows the functions computed by the four gates so far considered.

The NAND and NOR gates have the advantage that a circuit to carry out any binary logical function can be constructed using either NAND gates only, or NOR gates only. As the table shows the output from a NAND gate is the complement of the output from an AND gate. In other words the AND gate computes the function AB while the NAND gate computes the comple-

Inputs		Gate output function			
A	B	AND	OR	NAND	NOR
0	0	0	0	1	1
0	1	0	1	1	0
1	0	0	1	1	0
1	1	1	1	0	0
Function		AB	$A+B$	$\overline{AB} = \bar{A}+\bar{B}$	$\bar{A}+\bar{B} = \overline{AB}$

Table 1.1 Gate functions.

ment of the function, namely $\overline{AB}$ but from De Morgan's theorem this complement value $\overline{AB}$ is the same as $\bar{A}+\bar{B}$. Extensions of these functions for a situation involving more than two variables is elementary. The output of one gate can become the input to another so that more complicated functions can be computed by arranging the gates in a cascade or a tree-like manner.

Circuits, like those mentioned, where two or more inputs are combined to produce one or more outputs are termed combinational circuits. However many arithmetic functions involve a sequence of operations where one function after another has to be computed, these are known as sequential circuits and it is inherent that they should have a memory. In practice there are three types of circuit as defined below:

(a) Combinational circuits which have no memory.

(b) Memory circuits which just store and compute nothing.

(c) Sequential circuits which incorporate both memory and combinational circuits.

Even the most elementary operation involves many components within the computer and an important part of the design of the computer is the way in which it is organized to perform sequential operations.

The control unit contains most of this logic circuitry along with address registers to distribute information within the computer. In fact the control unit can be said to have two functions to perform; firstly to interpret the instruction supplied from the memory unit and secondly, on the basis of this interpretation, it must indicate to the arithmetic unit what it is to do. This latter function is achieved by electronic control signals. In accordance with these two functions the part of the control unit that interprets, or decodes, the instructions (instruction decoder) can be separated from the part that generates the control signals (control generator).

After the interpretation of an instruction from the instruction decoder, the control generator senses this interpretation and produces the necessary signals for the arithmetic unit. It also generates signals to choose the proper arguments (numbers) from the memory and sends them, when required, to the arithmetic unit. Other control signals take the result from the arithmetic unit back to the memory. After the execution of one instruction the control generator then produces signals to enable the next instruction to be brought from the memory to the instruction decoder. This illustrates the logical or sequential performance of instructions. The arithmetic unit itself contains shift registers to enable simple mathematical operations to be performed, results from the calculations being held in one of the registers known as the accumulator.

The input and output have for a long time been handled by electromechanical devices for reading and punching cards or tape, but it is then necessary to have intermediate devices for generating and interpreting this information for human users. It is now more commonplace for output to be generated on line printers, graph plotters or display units so that the results of a computation can be seen immediately (see also section 1.7). The same convenience of display for input is not commonly available although the use of teletypes connected in an interactive mode (see section 1.7) to computers is increasing. This forms a two-way slow communication between user and machine and represents the most elementary man–machine interaction.

Information is stored in the computer in the form of binary digits (bits), a computer 'word' consisting of, say, 24 bits. Fixed point numbers or integers are stored in this way, i.e. one for every word, but the range of numbers that can be stored is increased by using two or more words per number, for instance floating point or double precision numbers. Arithmetic is performed with both types of number, but it is faster with fixed point than with floating point, and consequently cheaper.

Storage within the machine is carried out on two main levels, the first being random-access store which allows any word to be read or written in a short time (1 μs is typical). This store is limited in size by its relatively high cost, the normal form being a ferrite core. The term random-access is applied since the information held in any one ferrite ring can be made available as fast as that from any other ring. In practice the cores are grouped into lines, each line being equivalent to a word (small machines may have 12, 16 or 18 bits per word, while large machines may use 36 bits). Each ring handles one bit of information, the ring being either in the 1 state or the 0 state, inferring that the magnetic flux flows either clockwise or anticlockwise round the ring. The principle of operation is based on the fact that when a current flows in a wire a magnetic field is set up around the wire. When the wire is surrounded by a ferrite ring the current should be large enough to permanently magnetize the ring in either direction. An even faster random access memory ($\sim$100 nanoseconds) uses semiconductor elements and differs in that it does not retain information when the machine is switched off.

The second level of storage is known as bulk store or sequential store, where words in sections of store are to be read or written sequentially. Examples of bulk storage are magnetic tapes or discs and drums. The magnetic tape is the cheapest form. Its main disadvantage is the fact that information stored in the middle of the tape can only be accessed by spinning the tape from the end. This relatively slow information access is largely removed by using magnetic discs or drums. A magnetic pickup can enter the disc or drum at any position to isolate particular information and these are becoming preferred on modern computers because they have a random-access time of milliseconds. As a summary to the discussion of different memory devices for the store, an approximate cost analysis per bit is shown in table 1.2. The particular choice for any computer obviously becomes a compromise between cost, speed and quantity needed.

	Cost per bit (p)	Access time
Magnetic tape	0.001	Seconds
Magnetic drum	0.01 }	Up to 25 milliseconds
Magnetic disc files	0.1 }	
Core stores	1	30 nanoseconds to 1 microsecond

Table 1.2 Approximate cost of storage memory devices.

1.5 Compilers and other computer software

Fortunately the input and output devices for digital computers are not limited to binary notation and most can transmit a reasonable number of symbols to the machine, although each is represented within the machine by some binary coding. This permits the languages adopted by the user in his programs to bear some resemblance to a living language, usually English. These user languages are known as higher level languages of which Fortran is one of the most universally adopted and for this reason will be used for all the examples to be covered in this book. The job of converting these languages into an acceptable form for the machine is achieved using a complex program known as the compiler which is usually kept in the memory of the computer.

When a program is read into the machine the compiler translates this program automatically into a machine language code, so relieving the user of the tedious job of converting the higher-level language instructions into a form acceptable to the computer.

As a system it can be seen that the computer is useless without its software to complement the hardware of the system. The compiler as part of the software provides a means by which the user is able to control the computer's behaviour in order to carry out the processes he requires.

As an example of the form which the machine language words take consider an I.C.L. 1900 series computer which has a word length of 24 bits. A machine language instruction is broken up within the word as indicated,

X	*F*	*M*	*N*
←3→	←7→	←2→	←12→

where X is one of the accumulators in which calculations are to be performed, F is the operator part of the word containing one of the different functions, M is the modifier which, depending on its value, will indicate whether the operand is to be used explicitly or is to be qualified:

$M = 0$ no modification

$\left.\begin{array}{l} M = 1 \\ M = 2 \\ M = 3 \end{array}\right\}$ modification by contents of relevant accumulator

N is the operand address which will identify a number on which to operate or a further instruction.

Within the memory, a data word is indistinguishable from an instruction. It is part of the control unit's task to select the next word for decoding as an instruction.

The program generated from the compiling process is often called the *object* program whereas the version, before translation, in the user's higher-level language is known as the *source* program. Sometimes for reasons of reducing computing time the object program can be permanently stored to avoid recompilation for every subsequent processing. Although different machines may possess different machine codes one benefit of automatic computing is that provided each machine has a compiler for some common higher-level language, then a program written in that language can be run on any one of the machines.

A schematic representation of the process of compiling and executing a Fortran program is shown in figure 1.3. The compiler program reads the

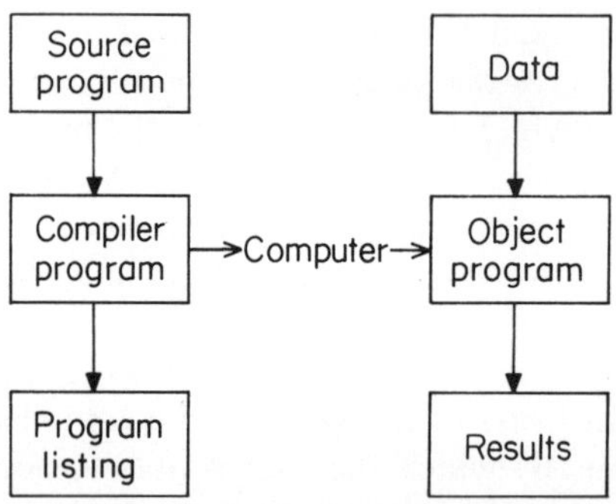

Figure 1.3 Compilation and execution.

source program supplied by the user (in a manner similar to the method outlined in the previous example), and produces the object program of machine code instructions. One of the outputs from this compilation process is usually the listing of the source program. In some machines the object program may be punched out on cards and then loaded into the computer.

In other systems, after compilation, the object program remains in the computer and control can be transferred automatically. The object program now reads the data and processes it according to the instructions generated from the source program. The results are printed on the listing or are stored on tape before the machine stops.

The languages (discussed in the following section) and the compilers associated with them form one part of the software system. Each statement in a language is translated into a larger number of simpler statements by the compiler. This process can be extrapolated so that a single statement can be used to achieve a complicated arithmetic process, such as solving a set of linear equations or evaluating a complex integral. These are examples of operations that are frequently required by users and such operations are written into subroutines which can be called into any program with the necessary calling statement. This system is known as a software subroutine library and the obvious advantage is that the availability of these 'standard' subroutines considerably reduces the programming effort required in producing a particular program. The subroutines available for any particular computer system vary and are influenced by the requirements of the users. However, most systems include subroutines for standard mathematical functions, statistical calculations and numerical analysis methods.

The software system plays an important part in the overseeing and general behaviour of the computing system. Various subsystems of this supervisory software ensure that the correct library routines are added to a program when required, the correct input/output devices are utilized, requests for any type of bulk auxiliary store are met and program error monitoring is carried out. This part of the software is also responsible for the processes involved in the compilation, assembly and execution of the program.

1.6 Programming languages

Programming, as we have seen, is the drawing up of schedules of instructions for a computer so that a specific task may be performed correctly. Nowadays there are hundreds of languages in 'common use' but it should be evident that the language forms the commonest method of communication between man and machine, and as such any language should be capable of easy understanding by both. A comparative study of programming languages must therefore be much more than a compilation of statements saying that one language allows a certain operation whereas another language does not. It must attempt to outline the principles of a 'good' language from a 'bad' one. Higman (1967) has suggested the following list of points to which a good language should adhere.

(a) Use a standard character set.

(b) Allow you, somehow or other, to do anything you want to do which can be defined in the context of your problem without reference to machine matters which lie outside this context.

(c) Make it easier to express the solution to your problem in terms which conform to the best practice in your subject.

(d) Allow such solutions to be expressed compactly without risking the obscurity which often accompanies compactness.

(e) Be free of any constructions which give rise to unintentional ambiguities.

(f) Permit the sort of ambiguity which is resolved dynamically, if required by the nature of the problem.

(g) Take over without change as much as possible of any well-formed descriptive language already established in the field in which the problem originated.

(h) Permit transcription, e.g. reading aloud, hand or typewriting as well as punched cards or tape.

(i) Allow of easy comprehension, i.e. it should be possible to read other people's programs more as one would a novel and less as one does an examination set book.

Complications arise with the development of any language and much of these can be attributed to the fact that some means of communicating are human and some are mechanical. This problem has led to the establishment of subsets or dialects of original programming languages which are peculiar to particular machines. A further difficulty can be introduced because of the variety of input/output devices available and the lack of consistency between them.

Basically, sentences (statements or instructions) consist of a common verb (operator), an object noun (operand) and an adverbial phrase or word (condition or modifier). The allowed relationships between the various words in a statement define the grammar of the language, and it is of importance to note that the relationships among words (precedence or order in which they occur) will effect the sense of a sentence as much as in English. Obviously to avoid tremendous complications of interpreting all English instructions, the number of command verbs and types of modifier must be severely restricted so that only a limited type of simple, but comprehensible, statement will be allowed. Two terms, syntax and semantics, are frequently used to describe the structure of a programming language and the way it is used. *Syntax*—the grammar, or relationships of one word to another is formalized, and the function of the compiler is to apply this formalism to translate into the basic machine code. *Semantics*—the meaning ascribed to the language in the context in which it is employed. The basic requirements for syntax and semantics are conflicting since, for maximum efficiency in the compiler, the allowed alternatives involved in the syntax should be kept to a minimum, implying that the language should be as simple as possible, whereas with the semantics maximum usage of the language requires that it should be as complex as possible. Obviously compromise is necessary, where the allowed alternatives in the syntax are limited to cope with as many situations as possible within a relatively restricted range.

For the present, and maybe for a long time to come, we must still use the restricted type of languages which have come to be known as problem-orientated or user-orientated. Although these two terms are not quite the same in meaning they both imply that the language has been devised to solve problems in a given class with some concessions to the convenience of the user. Fortran (FORmula TRANslater) is such a language, being very suitable for making arithmetic calculations from mathematical formulae. It exists in Marks I → V and in varieties such as Hartran and Madtran (in use at Harwell and M.I.T. respectively), the original language being developed for I.B.M. machines. Fortran has become the most universal language of its type and it is most often found in the Fortran II or IV version although small differences do occur in programming structure and facilities for different machines. The Fortran IV language profited from the development of Algol and includes a number of the features found in the latter, Algol (ALGOrithmic Language) being a similar problem-orientated language and existing in subsets such as Smalgol and Balgol ('Small Algol' and 'Burroughs Algol'). Cobol (COmmon Business Oriented Language) is an example of a language developed for particular types of problems of a commercial nature in applications such as invoices, pay-rolls and stock control. Its distinguishing feature, which it shares with a number of industrially sponsored languages such as the Honeywell 'Fact', the N.C.R. 'Language H', and the I.C.T. 'Nebuld', are that it attempts to be a subset of English and therefore readable by the uninitiated. It caters for data structures of the sort found in typical office records.

As a contrast to the problem-orientated languages mentioned there are also languages developed esentially for conversational mode operation (see next section). Examples of these types of languages are Focal (Formulating On-line Calculations in Algebraic Language), and Basic. The former exists specifically for use with Digital Equipment Corporation machines, e.g. PDP-8 range, while the latter is a more universally adopted language available for use with many different makes of machine. Both these languages consist of short imperative English statements which are relatively easy to learn. Mathematical expressions are typed, for the most part in standard notation, and because of their use in on-line conversational mode they provide an opportunity for initial computer programming from which some claim the graduation to use of languages such as Fortran becomes more straightforward.

Advanced languages at present are PL/1 (I.B.M.) and CPL (Cambridge/M.I.T.) both of which are far from fully operational. The former appears to be like a new Fortran with acknowledgements to Cobol while the latter resembles a new Algol language, but in both cases the sponsors have stressed that they have made a completely new start in the respective developments.

The objectives of the team using PL/1 (Programming Language no. 1) as set out by Radin and Rogoway (1965) were nothing to be illegal which makes clear and unambiguous sense and as far as possible it should never be necessary to escape into machine code. It should be modular, independent of machine characteristics, cater to the novice and be a programming rather than algorithmic language.

The aim in CPL (Combined Programming Language), as described in Barron and coauthors, was 'to produce a language which could be used for all types of problem, numerical and non-numerical, and would allow programmers to exploit all the facilities of a large and powerful computer without having to escape into machine code'.

Both these languages compare favourably with Higman's concept of a 'good' language.

Numerous other languages of considerable variety do exist for special purposes. Examples of such are 'string processing languages' (those with special features for statistical processes), simulation languages and the so-called 'open-ended' languages (possessing the property of describing possible changes in their own structure and then adopting them—as characteristic of natural languages). All of these are beyond the scope of this introduction and mention is only made for those who may wish to pursue an interest in the field of programming languages.

1.7 Man–machine interaction

There are two basic modes of computer operation, the most simple being the off-line batch processing mode. In this case the user has to prepare his input on cards or tape and submit them to a reception centre from which he will collect the results at some future date. This type of operation is obviously acceptable for such cases as, for example, the manipulation of experimental data, where the time between runs is long and the waiting time of processing and return of results is small in comparison. Often in such a situation the waiting time can be usefully employed in, say, obtaining a further set of data for the next run.

Batch processing, however, is less desirable in problems where modifications to, say, a design are dependent on results from a previous analysis. In this situation the waiting time incurred through batch mode is relatively useless since modification and progress cannot be made until the results are available. Another problem is that after each period of waiting time the designer has to remember what stage he had reached after the last processing and these breaks in continuity can often lead to a lower efficiency. For all these reasons an instantaneous response from the computer is desirable and this can only be obtained by a direct connection between the designer and the machine. This implies that the user must be in contact with and control of the machine during the whole of the processing. This effectively provides a definition of the second mode of operation, that of on-line or interactive mode processing.

The simplest form of communication is the teletype which is relatively cheap (≈ £700) and resembles in many ways the standard teleprinter. It can both punch and read paper tape giving in each case a hard copy on a paper roll. The most useful feature is that it can take signals corresponding to the paper tape input from a remotely connected line again giving a hard paper copy, and if this line is connected directly to the computer then information can be sent to and from the computer via the teletype over any distance. In the most elementary mode of operation the teletype can be likened to a very expensive adding machine when it is linked to the computer. It has the advantage of giving a complete record of everything done on the paper roll. Unfortunately the teletype is slow, the speed of typing being of the order of 10–15 characters per second and as such it has limitations in any computing application where this low speed is critical. It therefore becomes uneconomic and inefficient to use a teletype system when a large program or quantity of data is to be fed to the computer. The main uses of this type of communication are firstly in debugging (correcting) programs during development if the user has editing facilities to correct syntactic errors (and semantic errors if the program is run) through the keyboard. Secondly the teletype is useful for the interrogation, editing and updating of files (programs or data records) that are already stored. Thirdly it is of use for checking program segments, using limited data input to ensure that the results are consistent, before otherwise running the whole program and wasting computer time on an abortive analysis.

Some modifications can improve the low speed of the standard teletype, these being such peripheral devices as a fast tape reader (or card reader) on input and a line printer or graph plotter on output. The graph plotter in particular is used in engineering work where much of both the data and results can be expressed graphically. Simple graphic output from a computer, where a pen can move incrementally in any of four directions ($\pm x$ and $\pm y$), has been available for a long time and is in wide use. Refined versions are now available which overcome the problem of one-way transfer of information, inherent in the mechanical type of plotter, thus allowing the user to modify the visual output.

Graphic terminals, or computer-operated visual display units (VDU), basically consist of a display screen similar to a television screen incorporating a cathode ray tube where an electron beam is used to draw the picture rather than a pen as with the original mechanical plotters. The writing rate of these units is much higher than for the mechanical plotter and the information presented on the screen may be either in the form of diagrams or printed letters provided by the machine or by the operator through a keyboard. A keyboard input, a joystick spot position control or light pen, may be fitted to a VDU to provide computer input relevant to the display.

Of course, a teletype system can be made to print diagrams in a very laborious manner but the display units provide greater precision and a more visually pleasing form of output within a much faster time (up to 1 000 times faster). Their main disadvantage at the present time is cost which can be anything from two to several hundred times larger than for the simple teletype system. In the future it can be expected that the considerable work going into development of these units will eventually bring down costs as the demand increases and so make them more widely available. Often camera equipment for either moving or still pictures can be installed to record the displays or, as an alternative, instructions could be given to the computer to save the current screen display in a subsidiary display file.

1.8 Time sharing

In the early days of computing it was practical, though tedious, for a programmer to operate the computer on a personal basis, but as the machines became bigger, faster, more powerful and much more expensive this individual type of operation became totally impractical for large machines. Computer installations then incorporated software monitoring programs usually known as 'the operating system' which could supervise the operation of the machine.

Such systems generally handle the processing of users' programs sequentially and this is the principle of batch processing. Recognizing the need for interactive conversational processing for many engineering design problems and realizing that the response times of the designer are slow compared with the computer the computing resources should be used for other purposes while the user is thinking and modifying his particular program. This sort of sharing of computing facilities by several users is known as a 'time-sharing' mode of operation. With many time-sharing systems an order of priority exists for jobs and a low priority program may be interrupted during processing to allow a high priority program to be run. In most facilities, there will be a need for batch type operation as well, and hence many of the larger time-sharing systems include this mode. The batch jobs may be put into 'background status' and are only processed when an appropriate part of the system is idle because no calls for instantaneous service have been made.

A time-sharing system may allow the user to pay only for the time that he requires the machine. The computer system is centrally placed and each user is in contact via a remote terminal established through normal telephone connections. The terminal facilities are hired on an annual basis from the central computing installation and the cost is dependent on the facility required, i.e. teletype, graphics display unit, etc.

1.9 Computer classification

The most widely used method of classifying a particular computer system is probably that which is based on size and consequently cost. The size refers to the number of bits which can be internally stored in the machine and the cost includes all the hardware of the system. The cost of the individual machines is normally a function of the peripheral devices available as well as the size of the storage. Consequently at one end of the scale the so-called minicomputers (from £2 000 to £20 000) could have a few million bits of storage and include the standard accessories such as magnetic tape, teletypes, paper tape reader and punch. Moving up the scale the small (→ £100 000), medium (→ £250 000) and large (> £250 000) computers can have core sizes in excess of one million bits and include a much wider range of peripherals such as disc and drum storage, paper tape and card readers and punches, visual display units and high-speed line printers. The range of applications of these larger machines is vast and the increased adoption of installation networks provides even further storage capacity along with better reliability.

However, in terms of relative costs, the minicomputers are beginning to make an impact in that often their processing time for a particular program can be less than that of a larger machine. The restriction is on the core size, which may not affect many users and even large jobs can be processed by a skilful programmer adopting over-lay techniques. With development the costs of these minicomputer systems is becoming more competitive and in the future they may well be a better proposition for individual concerns in preference to a larger machine which will have to be on a time-sharing basis with several user groups so as to prove economical. A further advantage of these machines is that an owner can extend the facility by relatively small capital increments.

Notation	Definition	Section
F	operator of word	1.5
M	modifier of word	1.5
N	operand address of word	1.5
X	accumulator identifier of word	1.5
x	variable	1.2
y	variable	1.2

2 Introduction to Unit Computations

2.1 The process network

A chemical plant consists of equipment linked by pipe, etc, through which flows material for processing. It may be described by a network diagram in which the modules or nodes are units and the arrows represent streams. In any network all arrows may only be linked at a node so any junction of streams is shown as a unit on the diagram.

Input stream values are adjusted by the module to give the output stream values. To carry out a desired degree of change requires a separate input of information to each module. A byproduct of such calculation is a design rating of the unit.

The network for three modules is shown in figure 2.1 and it will be noted

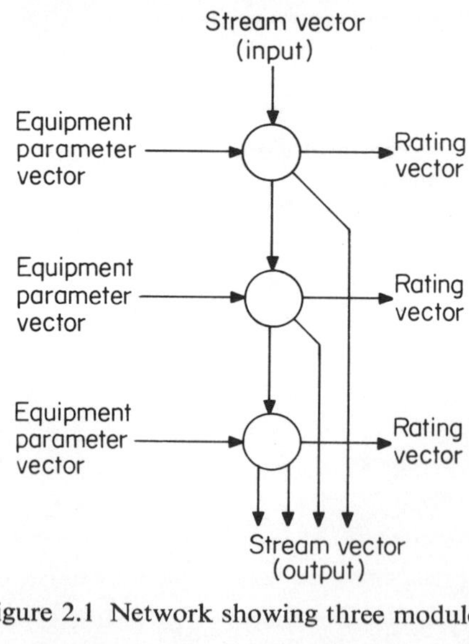

Figure 2.1 Network showing three modules.

that the equipment vectors for the three modules are not linked. Thus if the value of the input streams is known it is possible to carry out a unit computation for each module separately. Consider some of the simpler procedures used within these modules, such as the arithmetic operations used to modify stream values at junctions, reactors, heat exchangers, distillations, prime movers. *The degree of sophistication of the program required to carry out such computations depends on its current application, which in turn is based both on the required accuracy of financial information into which the calculations are eventually translated and the design costs involved in its use.* Classifying programs according to the stage of economic authorization is not possible because availability of both programs and data greatly influences the design costs, which are a function of labour and computing charges. However, it is possible to indicate an approximate classification according to strategy of use, i.e. by

Category A Preliminary balancing
Category B Short rating
Category C Full rating.

2.2 Category A Programs for preliminary balancing

These short programs do not give a design option and are esentially capable of carrying out mass balances over the unit, and setting temperature and pressure of output streams to preselected values. Normally the use of such programs is as subroutines in process simulators, the operation of which are described in chapter 6. However, as they are interesting as short computing exercises we will discuss them in some detail. Some typical modules are indicated in figure 2.2 with stream numbers applicable in the following formulae which use Fortran representation, arrays in capital letters, vectors in brackets.

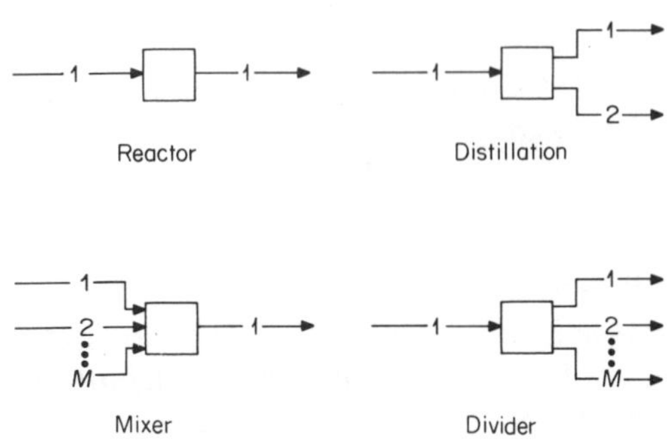

Figure 2.2 Input/output identification for some category A programs.

2.2.1 REACTION

The simplest form of reaction module involves input of a fixed percentage conversion, C, of a key component to products. Then if the stoichiometric factors $F(J)$ relative to this component are known (products positive, reactants negative), the stream composition for each component, J, can be calculated from

```
OUTPUT (1, J) = INPUT (1, J)+C * INPUT (1, K) * F(J)
```

It is essential using this formulation to have only one input and output stream, to identify the number of components, and from such an order the specific number of the key component, K.

2.2.2 DISTILLATION

A simple separation module can operate using a linear split of an input stream into the flows in the overhead and bottoms streams. The distribution factors $F(J)$ indicate the amount of a component in the feed which goes to the overhead stream, designated as output stream (1). Thus

```
OUTPUT (1, J) = INPUT (1, J)*F(J)
OUTPUT (2, J) = INPUT (1, J)-OUTPUT (1, J)
```

2.2.3 MIXER

To carry out only mass balances requires the identification of each component followed by the summation of all the input streams for each component. Thus for M input streams containing N components we can evaluate the system by

```
  DO 1 I = 1, M
  DO 1 J = 1, N
1 OUTPUT (1, J) = OUTPUT (1, J)+INPUT (I, J)
```

2.2.4 DIVIDER

Simple splitting of an input stream into a specified number of output streams, M, can be accomplished without change of composition by specifying the fraction of the input stream allocated to the Mth output stream

```
OUTPUT (M, J) = INPUT (1, J)*F(M)
```

2.2.5 PRESSURE AND TEMPERATURE ADJUSTMENT

For all the above systems it is possible to adjust temperature and pressure by either direct input of the desired value or as a default option setting them equal to that of the first input stream. This means that compressors and pumps need not be specified in this category. For the same reason it is probably best to define all heat exchanger programs as category B. This is because the calculation of enthalpy involves the input of physical property data which complicates the program. Also if such balances are required usually some form of rating of heat exchangers is necessary. It should again be noted that this is an extremely rough classification and, of course, does not imply that rating of equipment is essential for carrying out enthalpy balances.

2.3 Category B Programs for short rating

Such programs use short-cut methods of rating equipment which are sufficiently reliable to (with experience) size units to budget authorization accuracy ($\pm 15\%$), and which for optimization of process route permit achieving a sound optimum route for subsequent accurate evaluation at the control authorization stage. The use of these programs which should be short in execution time may be in a simulator (chapter 6) or as self-standing programs (chapter 4).

Where used for estimating purposes, ideally they should either generate the delivered cost of equipment (from correlation of, say, cost and heat transfer area for heat exchangers) or they should provide sufficient information for generation of costs in a separate costing program. Thus, for columns the number of trays, type, diameter and operating pressure should be specified; for exchangers surface area and operating pressure; for reactors wall thickness, volume, pressure, details of coils, stirrers, etc; for pumps the stages, horsepower, drive. A separate input of information about materials of construction is desirable.

The initial specification of common units which these programs provide may be sufficient in some circumstances to send to those manufacturers who provide a rating service using either category C programs or/and specialized know-how. Also they are useful for rating of existing units such as when debottlenecking or considering relocation of an item. They are invaluable for student design projects.

2.3.1 REACTION

Reactors may be designed based on equilibrium equations and for a specified temperature generating the conversion. The heat of reaction may be computed in a subroutine. Sizing is approximate using standard vessel geometry and frequently constant residence time. Alternatively a simple kinetic model might be included. Such programs tend to be specially written for specific problems. Franks (1972) has indicated several routines and discusses the mathematical modelling techniques involved.

2.3.2 DISTILLATION

The distillation model may be based on the rapid Fenske–Underwood–Gilliland correlation. This is detailed in chapter 4. Instead of specifying the distribution it is usual to give the composition of a product stream and identify the light and heavy key components. Normally, tray internals would not be considered and a typical output of the program is the feed position, diameter and number of trays, bubble and dew points of streams and reflux ratio, condenser and reboiler duties.

2.3.3 HEAT EXCHANGERS

In category B programs for heat exchangers it is usual to specify and assume correct an overall heat transfer coefficient, U. Otherwise it is necessary to carry out cyclic calculations until the geometry of the unit and this coefficient are in agreement. A common case is when input streams are fully specified, an approach temperature is given and the area is to be computed. A single tube pass, single shell case is assumed. The flowchart (figure 2.3) assumes the

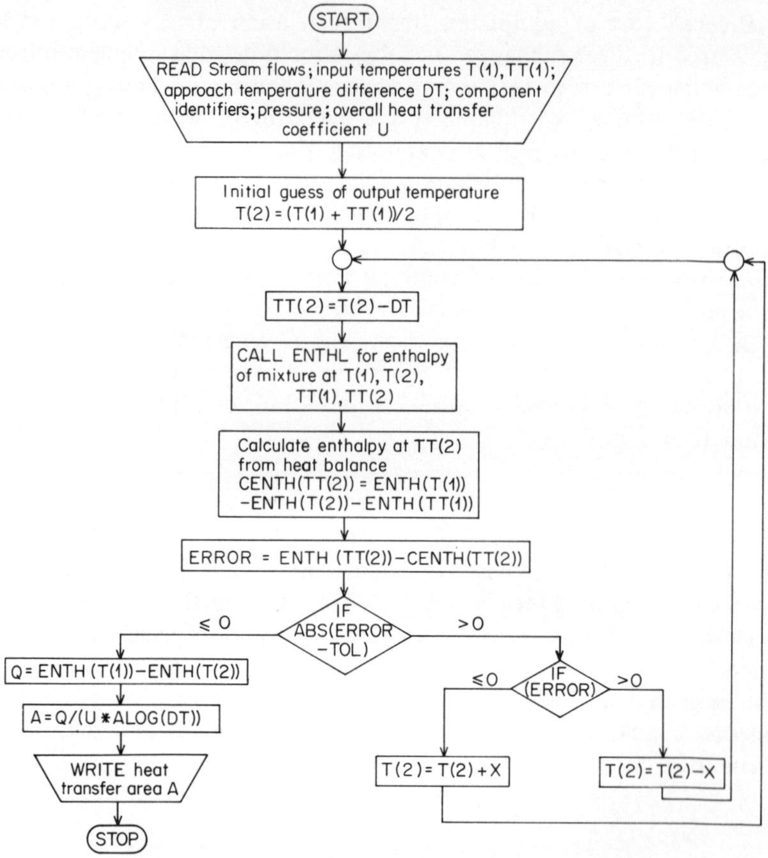

Figure 2.3 Flowchart for sizing simple exchanger.

availability of subroutines for evaluating enthalpy of the mixture and components. The main formulae are shown on the flowchart. Only a simple iterative method is used, convergence promotion being discussed later. A further case is when the heat transfer area is given and the output temperatures are calculated. Others arise in which assumptions may be made about the temperature of the outlet streams, i.e. cooling water outlet temperature, condensing a vapour, which again simplify the calculations.

2.3.4 MIXER

For the simple junction of streams without external stirring it is necessary to be able to compute enthalpies in order to carry out heat balances. Normally this is done by evaluating the enthalpy of the input streams separately by a suitable subroutine. These are then summed as per mass flow and the ensuing enthalpy value used in subroutines to evaluate the temperature of the output streams. The latter may be obtained by assuming a temperature, estimating the enthalpy, checking against the actual enthalpy and adjusting the guess accordingly. This is best done by using a process which converges rapidly to the correct value. Thus a variation of the Secant method, described in chapter 3, can be used by estimating as follows:

(a) Set iteration index *I*.

(b) Assume temperature.

(c) Calculate enthalpy for mixture at that temperature using physical data for each component.

(d) Define ERROR by

$$\text{ERROR} = \text{ENTHALPY (calculated)} - \text{ENTHALPY (actual)}.$$

(e) The iteration process for evaluating the temperature is then given by

$$\text{T(I+2)} = \text{T(I+1)} - \frac{\text{ERROR(I+1)*T(I+1)} - \text{T(I)}}{\text{ERROR(I+1)} - \text{ERROR(I)}}$$

(f) The calculation is stopped when the ERROR is within an acceptable limit, otherwise return to step (c).

2.3.5 COMPRESSORS AND PUMPS

For adjustment of pressure either a prime mover or valve is required. The latter is, of course, a simple adjustment to a preset value followed by a calculation to estimate any change in the vapour fraction of the stream.

For pumps, the desired output pressure, the motive power and estimated efficiency are normally supplied. Then total work, *W*, is given by

$$\text{W} = \text{F*M*V*P}$$

where *V* is molal volume
M is molar flow rate
P is pressure increase
F is factor for different units.

Single stage operation is normally assumed unless previous experience dictates to the contrary.

For compressors a knowledge of the number of stages to select is vital. The output pressure per stage is determined such that the work per stage is the same. Thus the pressure ratio $P(2)/P(1)$ is given by

P(2)/P(1) = (P(N)/P(1))**(1./(FLOAT(N)))

The total work, TW, is given by

TW = W*N

where W, the work per stage, is

W = (M*Z*R*T(1))* (K/(K−1))*((P(2)/P(1))**((K−1)/K)−1)

where Z is compressibility factor
R is gas constant
T is input temperature
K is specific heat ratio.

The problem for the designer is in establishing the correct number of stages before rating the unit. For this category of program the use of approximate guides as to the most suitable type of compressor for a given duty is satisfactory. Values of pressure/volume delivered are plotted in 'Capital Cost Estimation' (*Inst. Chem. Eng.*, 1969) and further indicate which type of compressor to select for a given duty. This graph is simplified in figure 2.4 in order to reduce the number of cases considered in student problems. The table is read in as four data arrays, VOL, PR, COST, STAGE into program 2.1. This actual subroutine is in fact part of a capital cost estimation program and the equipment parameters are read into the master segment. They have values as indicated in the comment segments. The program selects the number of stages and indicates the horse power and delivered equipment cost, DEC, of the compressor.

2.3.6 THERMODYNAMIC PACKAGE

Note how the availability of a physical properties/thermodynamic package simplifies the initial writing of this category of programs. We have found Motard's package in CHESS extremely useful for these programs, even when they are used in self-standing programs. This contains the properties of a 100 components measured as critical properties, molecular weight, acentric factor, molar volume, solubility parameter and coefficients for evaluating heat capacity ($a+bT+cT^2+dT^3$). Subroutines evaluate enthalpy, adiabatic flash, gas fugacity, K values, dew-point, bubble point, etc. The components must be identified for retrieval of the properties.

Problems can arise due to the bulk of such package. We prefer to reduce the size for a given problem by identifying the components required and accessing these properties from the data bank. The components and properties can then be relisted and used in a data bank prepared for the specific problem. The subroutines of the thermodynamic package are still used to evaluate values as they arise in the category B program but they are only incorporated when necessary and the COMMON statements are greatly modified.

If a further reduction in size is required it may be decided to incorporate the thermodynamic package in the program. For example, we have done this

KEY	£/BHP	TYPE
A	18	Single-stage vertical
B	11	Two-stage vertical
C	13	Two-stage vertical
D	15	Three-stage vertical
E	27	Two-stage vertical
F	66	Three-stage vertical
G	82	Four-stage vertical
H	7	Centrifugal

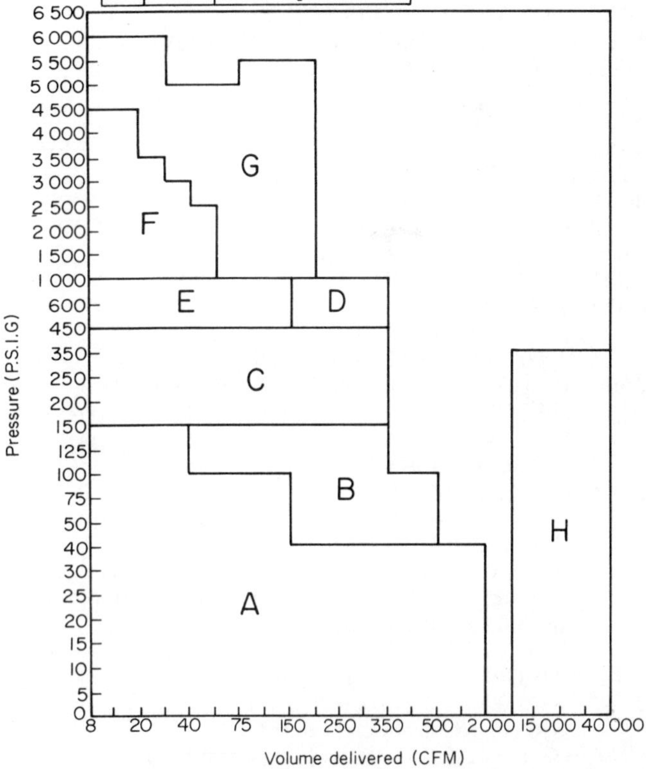

Figure 2.4 Prices for standard compressors

```
      SUBROUTINE COMPRESSOR
C     INDEX TO FACTORS FOR COMPRESSORS
C     UNIT(1) =          COMPRESSOR          =4
C     COST INDEX FACTORS
C     UNIT(2) =             1965
C     UNIT(3) =            1972
C     UNIT(4) =      VALUE OF HORSE POWER READ IN
C     UNIT(5) =      0 OR 1
C     UNIT(6) =      VOLUME DELIVERED IN CFM
C     UNIT(7) =      PRESSURE IN P.S.I.G.
C     UNIT(8) TO UNIT (10) EQUAL ZERO
C     UNIT(11) =     OUTLET PRESSURE
```

```
C  UNIT(12) =     INLET PRESSURE
C  UNIT(13) =     NO. OF STAGES
C  UNIT(14) =     INLET TEMPERATURE IN DEGREE R
C  UNIT(15) =     RATIO OF SPECIFIC HEATS
C  UBIT(16) =     VOLUME FLOW RATE OF FLUID
C  UNIT(17) =     EFFICIENCY OF COMPRESSOR
C  UNIT(18) TO UNIT (40) EQUAL ZERO
C  RECIPROCATING SINGLE AND MULTI COMPRESSORS AND
   CENTRIFUGAL
C  COMPRESSORS ARE COSTED ONLY
C  COMPRESSORS COST EXCLUSIVE OF MOTORS
C  SINGLE-STAGE VERTICAL COST £/BHP=£18       TYPE A
C  TWO-STAGE VERTICAL COST £/BHP=£11          TYPE B
C  TWO-STAGE VERTICAL COST £/BHP=£13          TYPE C
C  THREE-STAGE VERTICAL COST £/BHP=£15        TYPE D
C  TWO-STAGE VERTICAL COST £/BHP=£27          TYPE E
C  THREE-STAGE VERTICAL COST £/BHP=£66        TYPE F
C  FOUR-STAGE VERTICAL COST £/BHP=£82         TYPE G
   COMMON UNIT (40),FN(40),NT,DEC,DIC,INST
   DIMENSION ARR(20,29),VOL(20),PR(29),TYPE(20,29),V(20),PPR(29)
   DATA VOL/10.0,20.0,30.0,40.0,50.0,75.0,100.0,150.0,200.0,
  1          250.0,300.0,350.0,400.0,500.0,1000.0,2000.0,
  2          10000.0,15000.0,20000.0,40000.0/
   DATA PR/5.0,10.0,15.0,20.0,30.0,40.0,50.0,75.0,100.0,125.0,
  1          150.0,200.0,250.0,350.0,450.0,600.0,1000.0,1500.0,2000.0,
  2          2500.0,3000.0,3500.0,4000.0,4500.0,5000.0,5500.0,
  3          6000.0,6500.0/
   DATA ARR/15*18.0,0.0,4*7.0,15*18.0,0.0,4*7.0,
  1          15*18.0,0.0,4*7.0,15*18.0,0.0,4*7.0,
  2          15*18.0,0.0,4*7.0,15*18.0,0.0,4*7.0,
  1          15*18.0,0.0,4*7.0,8*18.0,5*11.0,3*0.0,4*7.0,
  1          8*18.0,5*11.0,3*0.0,4*7.0,8*18.0,5*11.0,3*0.0,4*7.0,
  1          4*18.0,7*11.0,5*0.0,4*7.0,4*18.0,7*11.0,5*0.0,4*7.0,
  1          11*13.0,5*0.0,4*7.0 11*13.0,5*0.0,4*7.0,
  1          11*13.0,5*0.0,4*7.0,11*13.0,9*0.0,
  1          8*27.0,3*15.0,9*0.0,8*27.0,3*15.0,9*0.0,
  1          5*66.0,4*82.0,11*0.0,5*66.0,4*82.0,11*0.0,
  1          5*66.0,4*82.0,11*0,0,4*66.0,5*82.0,11*0.0,
  1          3*66.0,6*82.0,11*0.0,2*66.0,7*82.0,11*0.0,
  1          2*66.0,7*82.0,11*0.0,9*82.0,11*0.0,
  1          3*82.0,3*0.0,3*82.0,11*0.0,3*82.0,17*0.0,
  1          20*0.0/
   DATA TYPE/15*1.0,0.0,4*1.0,15*1.0,0.0,4*1.0,
  1          15*1.0,0.0,4*1.0,15*1.0,0.0,4*1.0,
  1          15*1.0,0.0,4*1.0,15*1.0,0.0,4*1.0,
  1          15*1.0,0.0,4*1.0,8*1.0,5*2.0,3*0.0,4*1.0,
  1          8*1.0,5*2.0,3*0.0,4*1.0,8*1.0,5*2.0,3*0.0,4*1.0,
  1          4*1.0,7*2.0,5*0.0,4*1.0,4*1.0,7*2.0,5*0.0,4*1.0,
  1          11*2.0,5*0.0,4*1.0,11*2.0,5*0.0,4*1.0,
  1          11*2.0,5*0.0,4*1.0,11*2.0,9*0.0,
  1          8*2.0,3*3.0,9*0.0,8*2.0,3*3.0,9*0.0,
```

Program 2.1 Costing of a Compressor

```
      1         5*3.0,4*4.0,11*0.0,5*3.0,4*4.0,11*0.0,
      1         5*3.0,4*4.0,11*0.0,4*3.0,5*4.0,11*0.0,
      1         3*3.0,6*4.0,11*0.0,2*3.0,7*4.0,11*0.0,
      1         2*3.0,7*4.0,11*0.0,9*4.0,11*0.0,
      1         3*4.0,3*0.0,3*4.0,11*0.0,3*4.0,17*0.0,
      1         20*0.0/
      IF(UNIT(6).LT.8.0.OR.UNIT(6).GT.40000) GO TO 81
      IF(UNIT(7).LT.0.0.OR.UNIT(7).GT.6500) GO TO 83
      DO 40 I=1,20
      IF(UNIT(6).LT.VOL(I)) GO TO 35
   40 CONTINUE
   35 IX=I
      DO 60 I=1, 29
      IF(UNIT(7).LT.PR(I))GO TO 70
   60 CONTINUE
   70 IY=I
      WRITE(3,313)IX,IY
  313 FORMAT(2I3)
      RES=ARR(IX,IY)
      STAGE=TYPE(IX,IY)
      WRITE(3,75)RES
      WRITE (3,76)STAGE
   75 FORMAT(1H0,'COST OF COMPRESSOR IN £/BHP=',F12.4)
   76 FORMAT (1H ,'NO. OF STAGES=',F5.1)
      UNIT(13)=STAGE
      IF(UNIT(5).GT.0.0)GO TO 315
      P=(UNIT(11)/UNIT(12))**(1.0/UNIT(13))
      HP=0.0643*UNIT(13)*UNIT(14)*UNIT(15)*UNIT(16)*(P**(1.0-1.0/
     1UNIT(15))-1.0)
     1    /(520*(UNIT(15)-1.0)*UNIT(17))
      WRITE(3,301)HP
  301 FORMAT(1H0,'HORSE POWER=',F12.4)
      FIND=UNIT(2)*1.0+UNIT(3)*1.5
      DEC=RES*HP*FIND
      GO TO 316
  315 DEC=RES*UNIT(4)*1.5
  316 WRITE(3,80)DEC
   80 FORMAT(1H0,'DELIVER EQUIPMENT COST OF COMPRESSOR=
     1£',12.4) GO TO 310
   81 WRITE(3,82)UNIT(6)
   82 FORMAT(1H0,'VOLUME DELIVERED OUT OF CONSIDERATION
     1=',F12.4)
   83 WRITE(3,84)UNIT(7)
   84 FORMAT (1H0,'PRESSURE OUT OF RANGE CONSIDERED=
     1',F12.4)
      IF (INST) 311,311,310
  310 CALL INSTALL
      WRITE (3,312)DIC
  312 FORMAT (1H0,'INSTALLED COST OF COMPRESSORS=£',F12.4)
  311 CONTINUE
      RETURN
      END
```

Program 2.1 Costing of a Compressor—*continued*

in chapter 4 for explanatory purposes. In this case it is still useful to have a data bank to readily access the properties required.

These remarks also apply to category C programs whenever such information is required.

2.4 Category C Program for full rating

Category C programs are frequently used following evaluation of heat and mass balances and short-cut rating of equipment. They are therefore used largely as self-standing programs. Other factors such as computing charges, data requirement, specialized use and availability generally mean they are used after Budget authorization of capital expenditure. It is not possible to itemize a program of this nature because of lack of space to explain both theory and computational sequence, but we can indicate the philosophy of design pertinent to their construction and the criteria for use.

The early development of these programs involved the translation of time-honoured manual calculations into computer language. This resulted in the increased use of some procedures which although accepted as desirable methods of calculation were too complex and time-consuming for manual calculation. It was then appreciated that completely new approaches could be used, not only in the method of calculation adopted in order to obtain an answer which agrees with a set of data, but to attempt a mathematical simulation of the physical/chemical processes involved. The previous custom involved defining the appropriate dimensionless groups and determining the coefficient from experimental results. Now as an alternative solution it is often feasible to define the process as far as possible in terms of the fundamental relations for heat, mass and momentum transfer, provide estimates for unknowns and adjust these functions until the model describes the actual process as indicated by the data.

Such advanced models operating on the limit of the technology they represent are only used if they result in economic advantage. The model may be difficult to write and involve considerable time both in formulation and on the computer. Even when proven, problems arise due to the cost of running and the difficulty in obtaining the correct input data. On the other hand, before the availability of these programs many units could not be adequately designed, sometimes resulting in the process itself having to be altered. A further bonus is the reduction of capital cost of the unit as safety factors may be reduced or eliminated. Taborek and Palen (1972) indicate application of this philosophy in heat exchanger design.

Another advantage of using such computer-aided design of units is that on-stream time may be increased by the improved selection from different designs and the analysis of process reliability. However, *blind use of a computer model is wrong*. A personal experience was with respect to a process being commissioned with great difficulty because the condensers of two of the columns would not condense sufficient overhead vapour. The previous practice within the Company had been to condense these vapours on the shell side but the new units designed by process licensor reversed this procedure. The designs were calculated by computer and the belief of infallibility (prevalent in the days before computerized bills became a familiar everyday source of error) pervaded even when hand calculation indicated that the overall heat transfer coefficient was lower. Reasons were postulated that fouling on the cooling water side was the cause, a theory only disproved by cutting a hole in the shell and physically examining the outside of the tubes.

This illustrates how the engineering feel for the problem must be maintained. The use of sophisticated models should preferably be compared against actual performance of units before acceptance. It is essential that the design engineer uses the models to aid his judgement rather than accepts them as automatic problem solvers. It is impossible to use such programs properly without having some idea of what they are attempting to do. Note, too, that any computer error is almost invariably a programming error rather than a hardware fault. Meticulous checking of any routine is essential.

2.5 General

This brief survey indicates the range of programs for rating units. The programs illustrated are of extremely simple procedures and give an introduction to writing process engineering-type programs if taken further by the reader. It is necessary before considering other programs to review some numerical methods which reduce the time required to achieve a solution. Following this we will return to the study of unit computations in chapter 4 with the unit Distillation as our example.

It should also be noted that each unit is part of a system and an optimum solution for the size of unit may only be achieved in the context of the overall scheme. This makes rough optimization of process important before making detailed design.

Notation	Definition	Section
A	heat exchanger area	2.3.3
a, b, c, d	heat capacity polynominal coefficients	2.3.6
C	percentage conversion	2.2.1
F	unit factor	2.3.5
$F(J)$	distribution factor	2.2.2
$F(J)$	stoichiometric factor	2.2.1
I	iteration index	2.3.4
J	component	general
K	key component	2.2.1
k	specific heat ratio	2.3.5
M	molar flowrate	2.3.5
M	number of input or output streams	2.2

N	number of components	2.2
N	number of compressor stages	2.3.5
P	pressure	general
Q	heat load	2.3.3
R	gas constant	2.3.5
T	temperature	general
T(1), *TT*(1)	heat exchanger streams—input temperatures	2.3.3
T(2), *TT*(2)	heat exchanger streams—output temperatures	2.3.3
TW	total work	2.3.5
U	overall heat transfer coefficient	2.3.3
V	molal volume	2.3.5
W	work	2.3.5
Z	compressibility factor	2.3.5

3 Some Numerical and Optimization Methods

Before describing a unit computation in greater detail it is desirable to indicate some numerical methods which are used in the subsequent examples. The particular methods may be used to further the programming knowledge of the reader if they are used to solve the particular mathematical problems they represent.

The areas covered are represented by the section headings:

3.1 Roots of equations
3.2 Simultaneous linear equations
3.3 Optimization techniques
3.4 Correlation of data

and further information is obtained in other chapters in which they are applied. These methods may according to circumstances be used either as subroutines into which information is sent from the unit procedure programs or be fully integrated within these programs.

3.1 Roots of equations

Often engineering problems require, at some stage, determination of the roots of an equation defined as the values of x which satisfy an equation of the form,

$$f(x) = 0$$

Examples of such equations are

$$4x^3+3x-12 = 0 \text{ polynomial}$$
$$\cos x - x = 0 \text{ transcendental}$$

Values for real roots of an equation may be obtained by graphical solution, i.e. using a graph plotter on an interactive computer facility. A further simple method of solution is that of trial and error which involves repetitive substitutions using an incremented value of the variable concerned for each successive iteration. It can produce an accurate solution if the increment is small but it is a very time-consuming and hence expensive method of isolating the required value.

Other methods generally start with an initial estimate and calculate successive trials by some analytical technique as will be indicated.

3.1.1 PROPERTIES OF POLYNOMIALS AND TRANSCENDENTAL FUNCTIONS

An nth order polynomial is of the form

$$f(x) = a_1x^n + a_2x^{n-1} + a_3x^{n-2} \dots + a_nx + a_{n+1}$$

where $a_1 \neq 0$. When the polymonial is equated to zero there must be n roots which will satisfy the equation. The roots may be either real or complex but if all the coefficients a_i are real then any complex roots must occur in pairs or complex conjugates. The roots may also be either distinct of non-distinct (the latter implies that there is more than one root with the same value). Examples illustrating these points are:

$x^2+2x-3 = 0$ has real roots $x = 1$; $x = -3$
$x^2+4x+8 = 0$ has complex conjugate roots $x = -2+2i$; $x = -2-2i$
where $i = \sqrt{-1}$
Both these equations show distinct roots.
$x^2-2x+1 = 0$ has non-distinct roots $x = 1$; $x = 1$
The root $x = 1$ has a multiplicity of 2.

A transcendental function is defined as one whose value cannot be determined for any given argument by a finite number of additions, subtractions, multiplications, divisions and extractions of roots. Examples are trigonometric, logarithmic, exponential and hyperbolic functions. Because of their many various forms general properties are difficult to define. They can possess both real and imaginary roots and some may have a finite number of real roots while others may have an infinite number. The number and locations of the real roots are probably best found using graphical techniques, although methods of approximate conversion to polynomial equations do exist. Examples of transcendental equations are

$$x^2 - \tan x = 0 \quad \text{and} \quad xe^x = 0$$

3.1.2 HALF-INTERVAL METHOD

A root of an equation can be located using this method by evaluating the function at the extremes and midpoint of a subinterval which is known to contain a root. One method of defining this subinterval is to apply a crude search (i.e where the increment of the variable is large for each subsequent function evaluation) to show where the function value changes sign. This indicates that the root of the equation must lie within this region since the function therein must have a zero value. Thus if three function values are determined they must show a change in sign, and the half-interval which contains this sign difference is retained while the other half-interval is discarded. In this way each iteration reduces the subinterval size by a factor of a half and eventually the value of the root is isolated when the function equals zero. Note that the number of trials required to isolate the root is reduced according to the error acceptable in the value of the root.

An illustration of the use of this method is outlined in table 3.1. Suppose that the initial search shows that a root of the equation

$$x^3-2x^2-19x+20 = 0$$

lies within the subinterval $x = 2.0$; $x = 10.0$.At $x = 6.0$ the function value

x_a	x_{mid}	x_b	$f(x_a)$	$f(x_{mid})$	$f(x_b)$
2.0	6.0	10.0	−18.0	+50.0	+630.0
2.0	4.0	6.0	−18.0	−24.0	+50.0
4.0	5.0	6.0	−24.0	0.0	+50.0

Table 3.1 Half-interval method.

is +50.0 therefore the root must lie between $x = 2.0$ and $x = 6.0$ (i.e. within the half-interval exhibiting the sign change). Three initial function evaluations are made followed by one further calculation for each step afterwards, and as can be seen from the table the root of the equation is isolated at $x = 5.0$ after three steps.

The other two roots can be found by reducing the polynomial equation to a quadratic, with division by a factor $(x-5)$, and then applying the standard quadratic formula

$$x = \frac{-b\pm\sqrt{b^2-4ac}}{2a}$$

Problems are encountered in this search method when the interval chosen to isolate the subinterval to be explored is not large enough to contain all or any of the roots, and conversely the search may fail if two roots are close together and the increment of interval is too large.

3.1.3 SECANT METHOD

This method of finding a root of an equation may best be illustrated by reference to figure 3.1, where x_1 and x_2 represent two arbitrary initial values, and $f(x_1)$ and $f(x_2)$ represent their respective function values. The third trial value x_3 is taken as point C on the line ABC. The fourth point x_4 is similarly found by using the points x_2 and x_3; and the method continued until the root is obtained within a desired accuracy.

Again considering the equation

$$f(x) = x^3-2x^2-19x+20 = 0$$

using $x = 10.0$ as the first trial gives $f(10) = 630.0$, while a second trial of $x_2 = 8.0$ gives $f(8.0) = 252.0$.

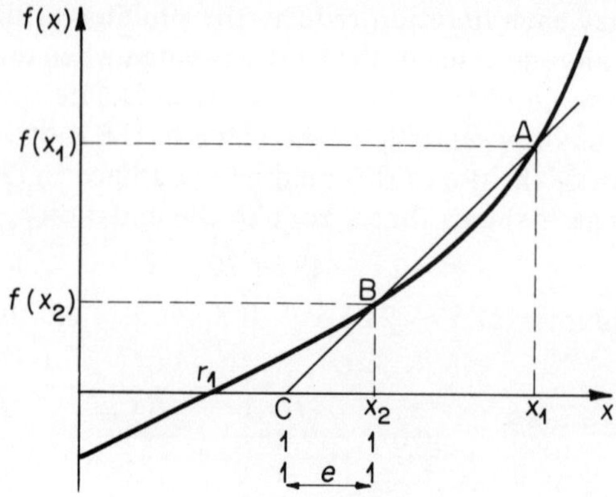

Figure 3.1 Secant method

Referring to figure 3.1, the value of e from similar triangles is

$$\frac{e}{f(x_2)} = \frac{e+(x_1-x_2)}{f(x_1)}$$

i.e.

$$\frac{e}{252.0} = \frac{e+(10-8)}{630.0}$$

$\therefore$

$$e = 1.33$$

and $x_3 = x_2 - e = 8.0 - 1.33 = 6.66$.

This process is continued as shown in table 3.2. The root value, r_1 is 5.0 and

x	$f(x)$	e
10.0	630.0	–
8.0	252.0	1.33
6.66	100.22	0.88
5.78	36.46	0.50
5.28	11.12	0.22
5.06	2.20	0.05
5.01	0.36	0.01
5.00	0.0	

Table 3.2 Secant method.

located after eight trials. An adaptation of this method is known as linear interpolation or the method of false position. The first two trials are chosen so that the two functional values are of opposite sign. The equation root lying between these two points can then be approximately obtained by linear interpolation (see figure 3.2).

Thus

$$\frac{e}{f(x_1)} = \frac{x_2-x_1-e}{f(x_2)}$$

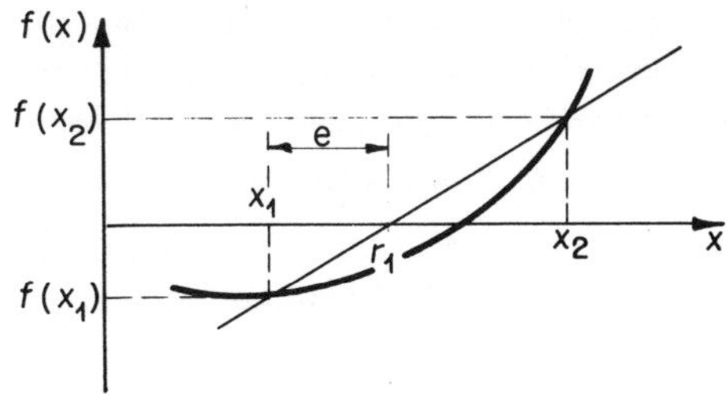

Figure 3.2 False position method

As an example consider the equation,

$$x - \cos x = 0$$

An initial search shows that the function value changes sign in the subinterval of $x = 0.6$ and $x = 0.8$. The solution of the root of the equation is found after four trials as tabulated in table 3.3. A particular case where the Secant

x	$f(x)$	e
0.6	−0.225	
0.8	+0.103	0.137
0.737	−0.0035	0.002
0.739	0.0	

Table 3.3 False position method.

method is not applicable is when $f(x_1) \approx f(x_2)$ since this produces a third trial point near infinity. The method of false position eliminates this particular problem but also has the limitation that it does not converge for functions which are tangential to the x axis.

3.1.4 NEWTON'S METHOD

Newton's method of determining a root of an equation is illustrated in figure 3.3. If x_1 represents some arbitrary first trial value then the second trial point,

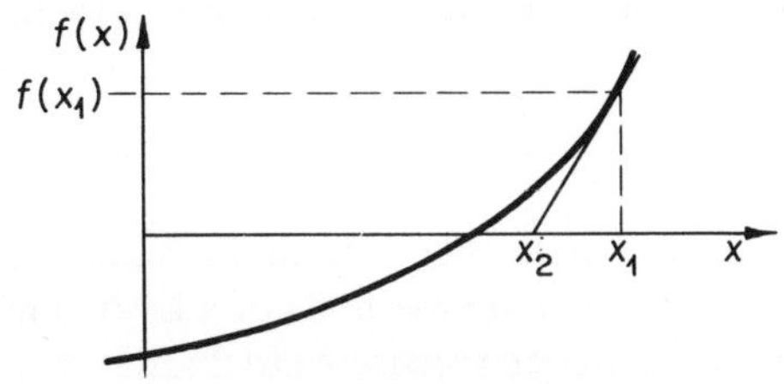

Figure 3.3 Newton's method.

x_2, is found by drawing the line tangent from the curve representing the function $f(x)$ at x_1. The intersection point on the x axis is taken as x_2. By definition of the tangent, the slope m of the tangent line is given by

$$m = f(x_1)/(x_1 - x_2)$$

Elementary calculus states that m is the first derivative of the function at point x_1, namely $f'(x_1)$. Hence the formula for locating the second trial point x_2 by Newton's method is

$$x_2 = x_1 - \frac{f(x_1)}{f'(x_1)}$$

As with the other methods a third trial is found from the second and so on until the root is located.

Again consider the equation

$$f(x) = x^3 - 2x^2 - 19x + 20 = 0$$
$$f'(x) = 3x^2 - 4x - 19$$

As before assume the first trial point $x_1 = 10.0$, then

$f(10) = 630$

$f'(10) = 241$

$\therefore \quad x_2 = 10.0 - 630/241 = 7.39$

Table 3.4 indicates how the value $r_1 = 5.0$ is found by successive trials.

x	$f(x)$	$f'(x)$	$f(x)/f'(x)$
10.0	630.0	241.0	2.61
7.39	173.95	115.28	1.51
5.88	42.43	61.203	0.69
5.19	7.316	41.050	0.18
5.01	0.36	36.26	0.01
5.00	0		

Table 3.4 Newton's method.

Newton's method does not always converge, for instance when the slope $f'(x_1)$ is close to zero since the next trial point x_2 will be near infinity. Also infinite looping may occur between trials x_1 and x_2 implying that the intersection of the tangent line from $f(x_1)$ with the x axis gives a point x_2 which by a similar construction of the slope from $f(x_2)$ yields $x_3 = x_1$; and then $x_4 = x_2$, etc. These peculiarities may be avoided sometimes by selecting a new trial to replace x_n.

3.1.5 DISCUSSION OF METHODS

The methods are generally applicable to both polynomial and transcendental equations and they are all iteration methods of a kind. The examples used to illustrate the methods have all been restricted to finding real roots of equations. All of these methods can be used to determine imaginary roots with the condition that complex arithmetic must be employed; this infers that the roots exist as complex conjugate pairs.

It can be shown by comparison of the rate of decrease of the error term for each trial that Newton's method will converge faster than the interpolation method and usually a little faster than the Secant method. The linear interpolation method has the disadvantage of not being able to determine even ordered multiple roots (e.g. $x = 1$; $x = 1$), although it will converge for all other types of roots whereas both the Secant and Newton's method might diverge. The half interval method tends to be similar to the interpolation method but is usually subject to slower convergence. Many other methods exist for solution of the roots of equations and all have their advantages and limitations. As an example, consider the iteration method of Richmond which is an extension of Newton's method. Here subsequent trial points are found according to the formula,

$$x_{n+1} = x_n - \frac{2f(x_n)f'(x_n)}{2(f'(x_n))^2 - f(x_n)f''(x_n)}$$

This method requires the evaluation of $f(x)$, $f'(x)$ and $f''(x)$ at each iteration step. The improved rate of convergence attained must be balanced against the time required to calculate $f''(x)$, which is not required in Newton's method, but more importantly the method assumes that both $f'(x)$ and $f''(x)$ can be evaluated at each step.

The rate of convergence is an important parameter for any iteration process and it is possible with many methods to apply acceleration techniques to promote faster convergence. One such method is that of Wegstein which when applied tends to speed up convergence and indeed can even force convergence on a system where poor initial choice would normally not permit convergence to be reached. Note, of course, that this method does not guarantee convergence.

The main feature of this technique is based upon the use

$$\bar{x}_{n+1} = qx_n + (1-q)x_{n+1}$$

x_n and x_{n+1} are found from the normal iteration cycle and then the previous equation is applied, with a specific value of q, to calculate $\bar{x}_{n+1}$ which is a better value than x_{n+1} and so is used in preference to generate x_{n+2}. The main problem is to assign a realistic value to q which will attain the required result. Unfortunately q is dependent on $\bar{x}_r$, the unknown root value which is to be isolated. With the aid of a graphical analysis an approximate value of q can be found from

$$q = \frac{b}{1-b}$$

where $b \approx -\dfrac{x_{n+1} - x_n}{x_{n-1} - x_n}$

The overall iteration process is listed by Lapidus (1962) as:

(a) Pick x_0.
(b) Calculate x_1 and x_2.
(c) Calculate b for x_0, x_1, x_2.
(d) Calculate q from b.
(e) Using q, x_1, x_2 calculate $\bar{x}_2$.
(f) $\bar{x}_2$ will be a closer approximation to $\bar{x}_r$ than x_2 so replace x_2 with $\bar{x}_2$.
(g) Using $\bar{x}_2$ generate new value x_3.
(h) Repeat entire process from step 3 using x_1, $\bar{x}_2$, x_3 to give b, then generate $\bar{x}_3$ to replace x_3, etc.

Effectively this method replaces the value $f'(x)$ required in Newton's method by a difference quotient defined in terms of previously calculated values (this refers to the variable b which is found from an approximation to the slope of the function and is therefore analogous to the derivative of the function). A similar expression is given by Crowe and coauthors (1971) where

$$x_{n+1} = x_n + k_s(x_n - x_{n-1})$$

and

$$k_s = \frac{x_n - x_{n-1}}{2x_{n-1} - x_n - x_{n-2}}.$$

The value of x_{n+1} is then used in the next cycle. This convergence promotion extrapolation is normally applied at every third or fourth iteration so that perturbations of the data can die out between applications.

3.2 Simultaneous linear equations

3.2.1 MATRIX INVERSION METHOD

When solving simultaneous linear equations it is preferable to use the shorthand nomenclature of matrices where the variable symbols are omitted since all the operations involve simple arithmetic on the coefficients and constants only. Thus consider the system of equations

$$\begin{aligned} a_{11}x_1 + a_{12}x_2 + \dots + a_{1n}x_n &= C_1 \\ a_{21}x_1 + a_{22}x_2 + \dots + a_{2n}x_n &= C_2 \\ \vdots \qquad \vdots \qquad\qquad \vdots \quad & \\ a_{n1}x_1 + a_{n2}x_2 + \dots + a_{nn}x_n &= C_n \end{aligned}$$

where the a_{ij}'s are known coefficients, the C_i's are known constants and the x_i's are unknowns for which the equations are to be solved. The matrix formed from these equations is $\boldsymbol{AX} = \boldsymbol{C}$

or

$$\begin{bmatrix} a_{11} & a_{12} & a_{13} & \dots & a_{1n} \\ a_{21} & a_{22} & a_{23} & \dots & a_{2n} \\ \vdots & & & & \\ a_{n1} & a_{n2} & a_{n3} & \dots & a_{nn} \end{bmatrix} \begin{bmatrix} x_1 \\ x_2 \\ \vdots \\ x_n \end{bmatrix} = \begin{bmatrix} C_1 \\ C_2 \\ \vdots \\ C_n \end{bmatrix}$$

This may be written as

$$X = A^{-1}C$$

where A^{-1} is the inverse of the coefficient matrix which can be determined from the definition

$$A^{-1} = \frac{\text{adj } A}{\det A} \quad \text{if } \det A \neq 0$$

adj A is the adjoint matrix and det A is the determinant of matrix A.

X is evaluated by multiplying the matrices A^{-1} and C together. Let us define these terms and illustrate the calculations involved in solving the set of equations,

$$\begin{aligned} 3x_1 - x_2 + x_3 \quad &= 10 \\ 2x_1 + 3x_2 - 2x_3 &= 8 \\ 3x_1 - 4x_2 + 4x_3 &= 13 \end{aligned}$$

The adjoint matrix is a transposed matrix of cofactors. Given a square matrix A of order n

$$A = \begin{bmatrix} a_{11} & a_{12} & \cdots & a_{1n} \\ a_{21} & a_{22} & \cdots & a_{2n} \\ \vdots & \vdots & & \vdots \\ a_{1} & a_{n2} & \cdots & a_{nn} \end{bmatrix}$$

the elements of the cofactor a_{11} are

$$\begin{bmatrix} a_{22} & \cdots & a_{2n} \\ \vdots & & \vdots \\ a_{n2} & \cdots & a_{nn} \end{bmatrix}$$

By definition the cofactor A_{ij} of a_{ij} in the matrix is found by multiplying the determinant obtained by deleting the ith row and jth column from the matrix by the term $[(-1)^{i+j}]$.

$$\text{adj } A = \begin{bmatrix} A_{11} & A_{21} & \cdots & A_{n1} \\ A_{12} & A_{22} & \cdots & A_{n2} \\ \vdots & \vdots & & \vdots \\ A_{1n} & A_{2n} & \cdots & A_{nn} \end{bmatrix}$$

Thus in the example stated

$$A = \begin{bmatrix} 3 & -1 & 1 \\ 2 & 3 & -2 \\ 3 & -4 & 4 \end{bmatrix} \quad \text{and } C = \begin{bmatrix} 10 \\ 8 \\ 13 \end{bmatrix}$$

The cofactor matrix elements are given by

$A_{11}=(1)(12-8)=4$ $\quad A_{12}=(-1)(8+6)=-14$ $\quad A_{13}=(1)(-8-9)=-17$
$A_{21}=(-1)(-4+4)=0$ $\quad A_{22}=(1)(12-3)=9$ $\quad A_{23}=(-1)(-12+3)=9$
$A_{31}=(1)(2-3)=-1$ $\quad A_{32}=(-1)(-6-2)=8$ $\quad A_{33}=(1)(9+2)=11$

These are transposed into the adjoint matrix,

$$\text{adj}\, \boldsymbol{A} = \begin{bmatrix} 4 & 0 & -1 \\ -14 & 9 & 8 \\ -17 & 9 & 11 \end{bmatrix}$$

The determinant of $\boldsymbol{A}$ is given by

$$\det \boldsymbol{A} = \sum_{\substack{j=1 \\ \text{for any } i}}^{n} a_{ij} A_{ij} = \sum_{\substack{i=1 \\ \text{for any } j}}^{n} a_{ij} A_{ij}$$

$$\det \boldsymbol{A} = (3)(4)+(-14)(-1)+(-17)(-1) = (3)(4)+(2)(0)+3(-1) = 9.$$

$$\boldsymbol{A}^{-1} = \begin{bmatrix} 4/9 & 0 & -1/9 \\ -14/9 & 1 & 8/9 \\ -17/9 & 1 & 11/9 \end{bmatrix}$$

note that if $\det \boldsymbol{A} = 0$ then $\boldsymbol{A}$ is said to be singular and $\boldsymbol{A}^{-1}$ does not exist.

To multiply the two matrices together they must be conformable, that is the number of columns of the first matrix must equal the number of rows of the second. In our example this is so.

$$\boldsymbol{X} = \boldsymbol{A}^{-1}\boldsymbol{C} = \begin{bmatrix} 4/9 & 0 & -1/9 \\ -14/9 & 1 & 8/9 \\ -17/9 & 1 & 11/9 \end{bmatrix} \begin{bmatrix} 10 \\ 8 \\ 13 \end{bmatrix}$$

and multiplying the rows of $\boldsymbol{A}^{-1}$ by the column of $\boldsymbol{C}$ gives

$$\begin{aligned} x_1 &= (\tfrac{40}{9}+0-\tfrac{13}{9}) = 3 \\ x_2 &= (^{-}\tfrac{140}{9}+\tfrac{72}{9}+\tfrac{104}{9}) = 4 \\ x_3 &= (^{-}\tfrac{170}{2}+\tfrac{72}{2}+\tfrac{143}{9}) = 5 \end{aligned}$$

One great advantage of this method is in the case of several systems of equations that have to be solved, all having the same coefficient matrix but different column vectors. In this situation an inverse of the matrix need only be calculated once, and only the multiplication operation must be repeated for each system. It is obvious that to solve a set of linear simultaneous equations using matrix methods the inverse of the coefficient matrix must be computed. The method of matrix inversion described previously involves the evaluation of the determinant of the matrix, an operation which can be ridiculously time consuming in a computer when the order of equations is perhaps ten or larger. Thus often for computer application the inverse of the matrix must be obtained by a different algorithm, one of which is closely related to the elimination techniques described in the next sections.

Basically the elimination method of matrix inversion involves forming an array of the elements of the coefficient matrix and an identity matrix of the same order as shown:

$$\begin{bmatrix} a_{11} & a_{12} & a_{13} & a_{14} & 1 & 0 & 0 & 0 \\ a_{21} & a_{22} & a_{23} & a_{24} & 0 & 1 & 0 & 0 \\ a_{31} & a_{32} & a_{33} & a_{34} & 0 & 0 & 1 & 0 \\ a_{41} & a_{42} & a_{43} & a_{44} & 0 & 0 & 0 & 1 \end{bmatrix}$$

Then by applying the elimination operations to be described, a matrix of the form illustrated is obtained:

$$\begin{bmatrix} 1 & 0 & 0 & 0 & \alpha_{11} & \alpha_{12} & \alpha_{13} & \alpha_{14} \\ 0 & 1 & 0 & 0 & \alpha_{21} & \alpha_{22} & \alpha_{23} & \alpha_{24} \\ 0 & 0 & 1 & 0 & \alpha_{31} & \alpha_{32} & \alpha_{33} & \alpha_{34} \\ 0 & 0 & 0 & 1 & \alpha_{41} & \alpha_{42} & \alpha_{43} & \alpha_{44} \end{bmatrix}$$

Here the right half of the matrix becomes the inverse of the left half of the original matrix. The allowable operations for this transformation by the elimination method, as we shall see, are:

(1) Multiply or divide any row by a constant.
(2) Add or subtract any row, element by element, to any other row.

3.2.2 GAUSSIAN ELIMINATION METHOD

Gaussian elimination or pivotal condensation is a method that is particularly suitable for computer solution and as such exists as a standard subroutine in many software libraries. Let us apply it to the set of equations:

(a) $x_1+x_2-2x_3 = 8$
(b) $2x_1-3x_2+x_3 = 5$
(c) $3x_1-2x_2+4x_3 = 6$

The two major steps involved are:

(1) Conversion of the coefficient matrix into a triangular matrix, a matrix in which the diagonal elements are not all zero and in which the triangle below the diagonal contains all zeros.
(2) Solution for x_n from x_1 by back substitution.

The first part of step 1 is to eliminate the coefficient of x_1 in (b) by multiplying (a) by a suitable constant and adding the product to (b). Similarly eliminate the coefficient of x_1 from (c) by multiplying (a) by some other suitable constant and adding the product to (c).

The augmented matrix becomes

$$\begin{bmatrix} 1 & 1 & -2 & 8 \\ 2 & -3 & 1 & 5 \\ 3 & -2 & 4 & 6 \end{bmatrix}$$

and the whole of step 1 is illustrated in the following operations. Subsequently in step 1 carry out the elimination of the coefficients of x_2, x_3, etc, by similar multiplication and addition calculation until the triangularization is complete.

Elimination of the coefficients of x_1

$$\begin{bmatrix} 1 & 1 & -2 & 8 \\ 2 & -3 & 1 & 5 \\ 3 & -2 & 4 & 6 \end{bmatrix} \begin{matrix} R_1/1 \\ R_2-2R_1' \\ R_3-3R_1' \end{matrix} \xrightarrow{} \begin{bmatrix} 1 & 1 & -2 & 8 \\ 0 & -5 & 5 & -11 \\ 0 & -5 & 10 & -18 \end{bmatrix}$$

Elimination of the coefficients of x_2

$$\begin{bmatrix} 1 & 1 & -2 & 8 \\ 0 & -5 & 5 & -11 \\ 0 & -5 & 10 & -18 \end{bmatrix} \quad \begin{matrix} \rightarrow \\ R_2/(-5) \\ R_3-(-5)R'_2 \end{matrix} \quad \begin{bmatrix} 1 & 1 & -2 & 8 \\ 0 & 1 & -1 & 1/5 \\ 0 & 0 & 5 & -7 \end{bmatrix}$$

In each part of step 1 the matrix is transformed by the indicated row operations. The row being used to modify the other rows (known often as the pivot row) is first divided by its diagonal element (i.e. a_{11}, a_{22}, etc, corresponding to R_1, R_2, etc). This ensures that the coefficient of the respective x_i that is being eliminated is unity, so avoiding the case when the multiplying constants would be fractions. The 'new' row created by this diagonal element division is marked with a prime.

The second step enables the value of x_n (here x_3) to be found and so back substitution yields the values of all x_i's. From the example,

solution of x_3

$$\begin{bmatrix} 1 & 1 & -2 & 8 \\ 0 & 1 & -1 & 11/5 \\ 0 & 0 & 5 & -7 \end{bmatrix} \quad \begin{matrix} \rightarrow \\ \\ R_3/(5) \end{matrix} \quad \begin{bmatrix} 1 & 1 & -2 & 8 \\ 0 & 1 & -1 & 11/5 \\ 0 & 0 & 1 & -7/5 \end{bmatrix}$$

and in equation form,

$$\begin{aligned} x_1+x_2-2x_3 &= 8 \\ x_2-x_3 &= 11/5 \\ x_3 &= -7/5 \end{aligned}$$

hence $\quad x_3 = -7/5 \quad x_2 = 4/5 \quad x_1 = 22/5$

It can be seen that this method produces a triangular matrix in that the final operation yields a coefficient matrix for which all the diagonal elements (a_{ii}) are 1 and all below them are zero. An alternative procedure to step 2, where back substitution is employed, is to continue the elimination technique so converting the coefficient matrix into a form in which only the diagonal elements are non-zero and the constant vector contains the values of x_i required. Referring to the previous example, forward elimination yielded

$$\begin{bmatrix} 1 & 1 & -2 & 8 \\ 0 & 1 & -1 & 11/5 \\ 0 & 0 & 1 & -7/5 \end{bmatrix}$$

Applying this approach gives the following.
Elimination of coefficients x_3

$$\begin{bmatrix} 1 & 1 & -2 & 8 \\ 0 & 1 & -1 & 11/5 \\ 0 & 0 & 1 & -7/5 \end{bmatrix} \quad \begin{matrix} \rightarrow \\ R_1-(-2)R_3 \\ R_2-(-1)R_3 \\ \\ \end{matrix} \quad \begin{bmatrix} 1 & 1 & 0 & 26/5 \\ 0 & 1 & 0 & 4/5 \\ 0 & 0 & 1 & -7/5 \end{bmatrix}$$

Elimination of coefficients x_2

$$\begin{bmatrix} 1 & 1 & 0 & 26/5 \\ 0 & 1 & 0 & 4/5 \\ 0 & 0 & 1 & -7/5 \end{bmatrix} \quad \begin{matrix} \rightarrow \\ R_1-R_2 \\ \\ \\ \end{matrix} \quad \begin{bmatrix} 1 & 0 & 0 & 22/5 \\ 0 & 1 & 0 & 4/5 \\ 0 & 0 & 1 & -7/5 \end{bmatrix}$$

The constant vector gives the solution

$$x_1 = 22/5 \qquad x_2 = 4/5 \qquad x_3 = -7/5$$

There are many other variations to this elimination method of linear simultaneous equations solution. One method is to omit the pivot row division by its diagonal element and include this factor in the multiplier, as stated before.

3.2.3 GAUSS–JORDAN METHOD

This method combines the forward and backward eliminations of the previous method, hence after one set of operation through the matrix the only elements remaining are the diagonal elements of 1 and the column vector solutions.

Consider again the previous example of a set of equations.

Elimination of coefficients of x_1

$$\begin{bmatrix} 1 & 1 & -2 & 8 \\ 2 & -3 & 1 & 5 \\ 3 & -2 & 4 & 6 \end{bmatrix} \begin{matrix} R_1/1 \\ R_2 - 2R_1' \\ R_3 - 3R_1' \end{matrix} \xrightarrow{} \begin{bmatrix} 1 & 1 & -2 & 8 \\ 0 & -5 & 5 & -11 \\ 0 & -5 & 10 & -18 \end{bmatrix}$$

Elimination of coefficients of x_2

$$\begin{bmatrix} 1 & 1 & -2 & 8 \\ 0 & -5 & 5 & -11 \\ 0 & -5 & 10 & -18 \end{bmatrix} \begin{matrix} R_1 - R_2' \\ R_2/(-5) \\ R_3 - (-5)R_2' \end{matrix} \xrightarrow{} \begin{bmatrix} 1 & 0 & -1 & 29/5 \\ 0 & 1 & -1 & 11/5 \\ 0 & 0 & 5 & -7 \end{bmatrix}$$

Elimination of coefficients of x_3

$$\begin{bmatrix} 1 & 0 & -1 & 29/5 \\ 0 & 1 & -1 & 11/5 \\ 0 & 0 & 5 & -7 \end{bmatrix} \begin{matrix} R_1 - (-1)R_3' \\ R_2 - (-1)R_3' \\ R_3/5 \end{matrix} \xrightarrow{} \begin{bmatrix} 1 & 0 & 0 & 22/5 \\ 0 & 1 & 0 & 4/5 \\ 0 & 0 & 1 & -7/5 \end{bmatrix}$$

Problems can arise in all these elimination methods when the diagonal elements (also known as pivot elements) are zero since one of the operations involves row division by these elements. This can often be avoided merely by rearranging the order of the equations within the matrix. The solution might depend on the order of equations if round-off errors are incurred (due to the fixed number of digits that can be retained for any number within a computer). Many techniques have now become available for incorporation into computer solutions to avoid inaccuracy produced from rounding errors and to select the 'best' way of presenting the equation within the matrix.

3.2.4 JACOBI ITERATION METHOD

This is probably the most simple iteration method and is also known as total step or complete iteration.

Consider the standard system of equations:

$$\begin{aligned} a_{11}x_1+a_{12}x_2+ \dots +a_{1n}x_n &= C_1 \\ a_{21}x_1+a_{22}x_2+ \dots +a_{2n}x_n &= C_2 \\ \vdots \qquad\qquad & \quad \vdots \\ a_{n1}x_1+a_{n2}x_2+ \dots +a_{nn}x_n &= C_n \end{aligned}$$

These may be rearranged so that the first can be solved for x_1, the second for x_2, etc.

$$\begin{aligned} x_1 &= \frac{1}{a_{11}}(C_1-a_{12}x_2-a_{13}x_3- \dots -a_{1n}x_n) \\ x_2 &= \frac{1}{a_{22}}(C_1-a_{21}x_1-a_{23}x_3- \dots -a_{2n}x_n) \\ &\vdots \\ x_n &= \frac{1}{a_{nn}}(C_n-a_{n1}x_1-a_{n2}x_2- \dots -a_{n,\,n-1}x_{n-1}) \end{aligned}$$

Initially the iteration starts with assumed values for $x_1^{(1)}$, $x_2^{(1)}$, ... $x_n^{(1)}$, and with these a new set $x_1^{(2)}$, $x_2^{(2)}$, ... $x_n^{(2)}$ can be calculated from the equations. The superscript represents the iteration number. The resubstitution of the calculated values $x_1^{(2)}$, $x_2^{(2)}$, ... $x_n^{(2)}$ yields a further set of values. This process is repeated until two successive sets of values differ by a margin less than the allowable deviation.

Let us again solve the system of equations:

$$\begin{aligned} 3x_1-x_2+x_3 \quad &= 10 \\ 2x_1-3x_2-2x_3 &= 8 \\ 3x_1-4x_2+4x_3 &= 13 \end{aligned}$$

Rearranging and adding an iteration number superscript, according to the method described, gives

$$\begin{aligned} x_1^{(r+1)} &= \tfrac{1}{3}(10+x_2^{(r)}-x_3^{(r)}) \\ x_2^{(r+1)} &= \tfrac{1}{3}(8-2x_1^{(r)}+2x_3^{(r)}) \\ x_3^{(r+1)} &= \tfrac{1}{4}(13-3x_1^{(r)}+4x_2^{(r)}) \end{aligned}$$

where r and $r+1$ are two successive iterations. Table 3.5 shows how the calculations proceed assuming the initial set of values to be $x_1^{(1)}=0$, $x_2^{(1)}=0$ and $x_3^{(1)}=0$.

The iteration techniques for the case of linear simultaneous equations solutions are more complicated than for a convergence problem involving only one variable since it is necessary to assume a complete set of values for the unknowns, and continue the process until all the unknowns have converged within the predefined allowable amount.

3.2.5 GAUSS–SEIDEL ITERATION METHOD

This method represents an improvement on the simple Jacobi iteration process since rather than using the entire set of values from the previous iteration in order to calculate the new set the most recently calculated variable at each

Iteration no. r	1	2	3	4	5	6	7	8	9	10	11	12	13	14
x_1	0	3.333	3.135	3.062	3.111	3.049	3.079	3.039	3.057	3.031	3.041	3.024	3.029	3.018
x_2	0	2.764	2.611	2.852	2.962	3.125	3.217	3.337	3.409	3.497	3.553	3.619	3.663	3.711
x_3	0	3.250	3.412	3.510	3.805	3.879	4.090	4.157	4.307	4.366	4.474	4.523	4.601	4.641

Iteration no. r	15	16	17	18	19	20	21	22	23	24	25	26	27
x_1	3.020	3.013	3.014	3.009	3.010	3.006	3.005	3.004	3.003	3.002	3.001	2.999	3.001
x_2	3.745	3.812	3.808	3.835	3.855	3.875	3.891	3.906	3.918	3.930	3.940	3.947	3.954
x_3	4.698	4.730	4.772	4.797	4.828	4.848	4.871	4.886	4.903	4.915	4.923	4.937	4.946

Iteration no. r	28	29	30	31	32	33	34	35	36	37	38	39	40
x_1	2.999	3.001	2.999	3.0	—	—	—	—	—	—	—	—	—
x_2	3.960	3.965	3.971	3.974	3.980	3.981	3.985	3.986	3.988	3.989	3.992	3.996	$3.998 \rightarrow 4.0$
x_3	4.953	4.961	4.966	4.971	4.975	4.979	4.983	4.986	4.988	4.991	4.992	4.995	$4.998 \rightarrow 5.0$

Table 3.5 Jacobi iteration method.

step is used. This difference might be best illustrated by considering again the general system of equations and comparing these with the system derived for the Jacobi iteration method.

$$x_1^{(r+1)} = \frac{1}{a_{11}}(C_1 - a_{12}x_2^{(r)} - a_{13}x_3^{(r)} - \ldots - a_{1n}x_n^{(r)})$$

$$x_2^{(r+1)} = \frac{1}{a_{22}}(C_2 - a_{21}x_1^{(r+1)} - a_{23}x_3^{(r)} - \ldots - a_{2n}x_n^{(r)})$$

$$\vdots$$

$$x_n^{(r+1)} = \frac{1}{a_{nn}}(C_n - a_{n1}x_1^{(r+1)} - a_{n2}x_2^{(r+1)} - \ldots - a_{n,n-1}x_{n-1}^{(r+1)})$$

Consider again the previous example. The progress of the iteration is shown in table 3.6 and it can be seen how this technique so reduces the required number of iterations.

Iteration no. r	1	2	3	4	5	6	7
x_1	0	3.333	3.079	3.017	3.001	2.999	2.999
x_2	0	0.446	1.412	2.221	2.802	3.199	3.467
x_3	0	1.199	2.352	3.209	3.801	4.201	4.469
Iteration no r	8	9	10	11	12	13	14
x_1	3.0	–	–	–	–	–	–
x_2	3.645	3.764	3.844	3.897	3.932	3.955	3.971
x_3	4.648	4.767	4.847	4.900	4.935	4.958	4.974
Iteration no r	15	16	17	18			
x_1	–	–	–	–			
x_2	3.981	3.988	3.993	3.999→4.0			
x_3	4.984	4.991	4.996	4.999→5.0			

Table 3.6 Guass-Seidel iteration.

These iteration methods may have the disadvantages of not converging or being slow to converge relative to the elimination methods except when a sparse coefficient matrix is encountered.

Diagonal dominance may be used as a check on whether the system is certain to converge. It exists if

$$|a_{pp}| \geqslant \sum_{\substack{q=1 \\ \text{but } q \neq p}}^{n} |a_{pq}| \qquad (p = 1, 2 \ldots, n)$$

and inequality for at least one value of p. For example

$$\begin{bmatrix} 4 & 1 & -1 & 1 \\ 1 & 3 & -1 & 1 \\ 0 & 2 & 4 & 2 \\ 0 & 0 & 2 & 3 \end{bmatrix} \quad \begin{array}{l} \text{i.e. } 4 > 3 \\ \quad 3 = 3 \text{ is diagonal dominant.} \\ \quad 4 = 4 \\ \quad 3 > 2 \end{array}$$

but

$$\begin{bmatrix} 2 & 1 & -1 & 1 \\ 1 & 2 & 1 & 0 \\ 0 & 0 & 4 & 1 \\ 1 & 1 & 1 & 3 \end{bmatrix} \quad \begin{array}{l} \text{i.e. } 2 < 3 \\ \quad 2 = 2 \text{ is not diagonal} \\ \quad 4 > 1 \text{ dominant in first row.} \\ \quad 3 = 3 \end{array}$$

However, note that the text example is not diagonal dominant but still converges. Thus this approach is solely a check. A better method is to calculate eigenvalues but this can in many cases take longer than simply trying out a particular set of equations.

3.3 Optimization Techniques

The mathematical model of a system for optimization may be written as

$$\max_{\boldsymbol{q}} \{\phi(\boldsymbol{p};\boldsymbol{q})|g(\boldsymbol{p};\boldsymbol{q}) \geqslant 0\} = \phi^*(\boldsymbol{p}) = \phi(\boldsymbol{p}, \boldsymbol{q}^*)$$

where ϕ is an objective function the value of which is determined by the decision variable vector $\boldsymbol{q}$ and the parameter vector $\boldsymbol{p}$. The various constraints on the system are indicated by the vector of constraint limitations $\boldsymbol{g}$ and may be classified as either equality or inequality in nature. Expressing the model in standard form permits the use of the appropriate optimization algorithm in a pack of subroutines, selection as to the appropriate method being made by the user. The correct algorithm permits adjustment of the decision variables $\boldsymbol{q}$ to obtain a best value of the objective function, ϕ^*, for the particular parameter vector $\boldsymbol{p}$ which describes the case being studied. Corresponding to ϕ^* there are one or more points, $\boldsymbol{q}^*$ where $\phi = \phi^*$. This approach is considered in more detail in subsequent chapters. Here we will consider simple equality functions.

3.3.1 OPTIMIZATION OF FUNCTIONS OF ONE VARIABLE

The general principle of direct search is to make a number of measurements of $y(x)$, the objective function, at values of x within a region $a \leqslant x \leqslant b$. At the end of the set of measurements hopefully the extreme value of the function will be in a certain subinterval of the region $(b-a)$. Such numerical approach introduces the concept of region elimination which may be used to isolate the optimum of a function that is unimodal.

Exhaustive search

Suppose a function, see figure 3.4, is defined on the interval (x_a, x_b), and has been evaluated at five equally spaced trial points within the region $x_a \rightarrow x_b$.

The function can be seen to have its largest value at x_2 and therefore the true maximum of the function must lie in the region $x_1 \rightarrow x_3$.

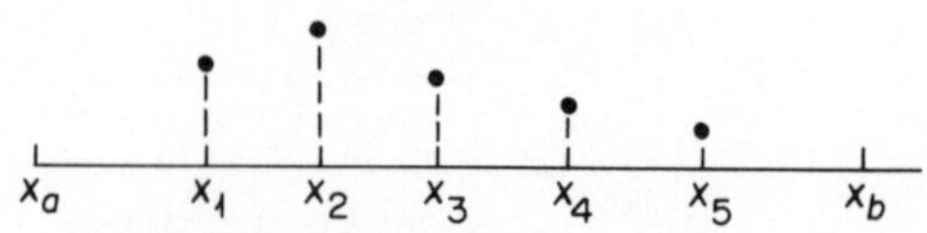

Figure 3.4 Exhaustive search.

After n trial points have been located within the overall interval of search of length I_0 and the function is found to be greatest at x_i then the new interval of search, I_1, must be defined by

$$I_1 = x_{i+1} - x_{i-1}$$

or in terms of the n trial points

$$I_1 = \frac{2I_0}{n+1}$$

The efficiency of the search procedure to computing time may be improved if information from the preceding trial is used to select the next trial. Such methods included Dichotomous Search, Fibonacci Search and the modified version of Fibonacci commonly known as the Golden Section. The latter method is extremely simple to program and will now be illustrated.

Golden section search

With the condition that the function is unimodal, this method involves placing two search points at a distance l from each side of the region of interest of length L_K, figure 3.5.

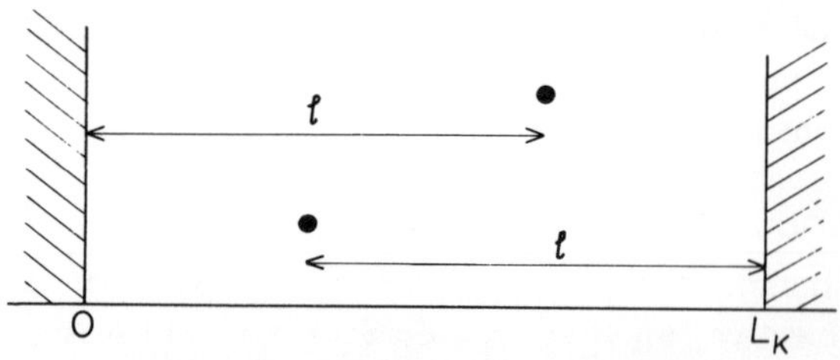

Figure 3.5 Initial location of points.

A section of the region may be eliminated by substituting these search point values into the objective function and comparing the two results. It is necessary to place these first two points so that the point remaining after the first region elimination is a fraction l of the distance from the side of the remaining region. Assuming $L_K = 1$, an approximate value of l can be derived, figure 3.6. Clearly $l^2 = 1 - l$ and hence $l = 0.618$.

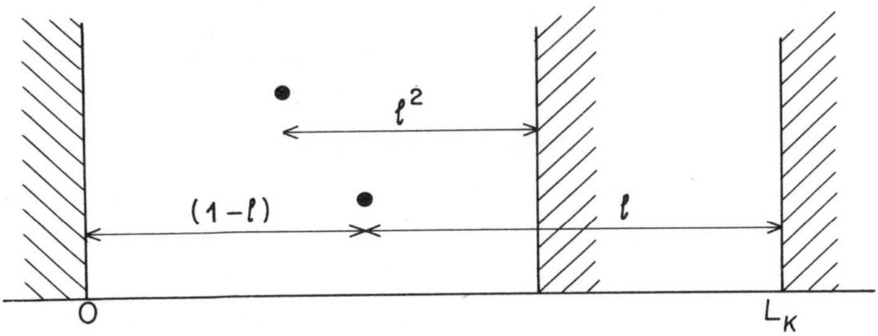

Figure 3.6 Application of region elimination.

A strict derivation of the above requires a study of the Fibonaccian series of numbers. Consider a region within the boundaries a and b, and having length L_K. We wish to place two search points $x_{1,1}$ and $x_{1,2}$ within the region so that $l = 0.618L_K$, where $L_K = b-a$.

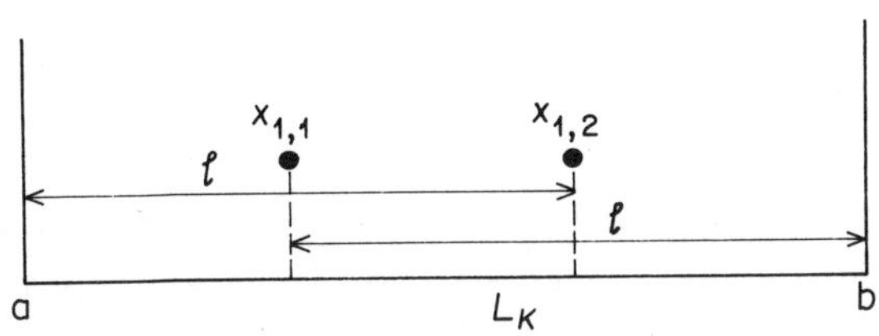

Figure 3.7 Definition of dimensions for Golden Section.

$$x_{1,1} = a+(L_K-l)$$
$$x_{1,2} = b-(L_K-l)$$

The system is resolved as follows. Assume we require a maximum (i.e. the value of x that yields the maximum value of the unimodal objective function, $y(x)$).

Cycle 1

Let $a = 0, \quad b = 1, \quad L_K = 1, \quad l = 0.618.$

$$x_{1,1} = 0+(1-0.618) = 0.382$$
$$x_{1,2} = 1-(1-0.618) = 0.618$$

If $y_{1,1}$ represents the objective function value for the point $x_{1,1}$, etc, and $y_{1,1} > y_{1,2}$, eliminate the region to the right of $x_{1,2}$.

Cycle 2

$$a = 0, \quad b = 0.618, \quad L_K = 0.618, \quad l = 0.618^2.$$
$$x_{2,1} = 0+(0.618-0.618^2) = 0.236$$
$$x_{2,2} = 0.618-(0.618-0.618^2) = 0.382$$

$x_{2,2}$ must be the same as $x_{1,1}$ since this is the point left from cycle 1 after the first region elimination. If $y_{2,2} > y_{2,1}$, eliminate the region to the left of $x_{2,1}$.

Cycle 3

$$a = 0.236, \quad b = 0.618, \quad L_K = 0.382, \quad l = 0.618\,(0.382)$$

$$x_{3,1} = 0.236+(0.382-0.236) = 0.382$$

$$x_{3,2} = 0.618-(0.382-0.236) = 0.472$$

As before, one value corresponds to a point from the previous cycle, i.e. $x_{3,1} = x_{2,2}$. If $y_{3,1} > y_{3,2}$, eliminate the region to the right of $x_{3,2}$ and so forth. The calculation continues until the region remaining is within some acceptable degree of accuracy. The optimum value of x is the final region boundary value giving the larger value of $y(x)$.

The same approach applies if a minimum function value is required, the only difference being the choice of region to be eliminated after each cycle.

As can be seen, this method of search rapidly reduces the potential range of values for the independent variable under examination, and an acceptable value is quickly isolated, the only limitation to this approach of search being that the function must be unimodal. A further example is given in chapter 4 and a suitable program is built into the Fenske–Underwood distillation program.

3.3.2 OPTIMIZATION OF MULTIVARIABLE FUNCTIONS—ANALYTICAL METHODS

Many optimization problems can be reduced by an appropriate series of eliminations to a problem involving only one independent variable but there are a number of instances when this elimination becomes either impossible or much too complex to be feasible. Hence, it is necessary to develop methods of optimization which can handle more than one independent variable.

The analytical methods of optimization may be classified according to the absence of constraints, the presence of equality constraints or the presence of inequality constraints.

In the first classification the maximum (or minimum) of an unrestricted objective function lies at a stationary point or at a position of discontinuity of the function or its derivative. This implies that we must evaluate the partial derivatives of the function and equate them to zero (calculus). However, if the partial differential equations do not provide for the solution of x_1, x_2, ... x_n, independent variables explicitly, iterative techniques for the solution of non-linear simultaneous equations may be required. Also in defining the regions in which differentiation of a continuous function may be carried out successfully, certain considerations must be made initially to avoid the presence of discontinuities in the objective function.

3.3.3 OPTIMIZATION OF MULTIVARIABLE FUNCTIONS—NUMERICAL METHODS

In many practical cases the analytical methods devised for optimization problems cannot effectively be applied because of either the mathematical operations required or the discontinuous behaviour of the functions involved.

Also it is unfortunate that the powerful search methods (such as the Golden Section, etc) for a one-dimensional function cannot be simply extended for multivariable function applications. So it is necessary to develop for multi-dimensional search entirely different approaches which are based on one central idea, namely, to locate a new feasible point which is an improvement on the existing value. This principle is applied iteratively until either no further improvement can be obtained or until the optimum is found within a desired degree of accuracy. Gradient methods of search have generally been found to be more effective than direct search methods but there are cases when the latter can be more appropriate, as when the function gradient is not analytically defined.

The range of direct search methods varies enormously from the most simple univariant search to the sophisticated acceleration in direction technique of Powell (1964) and the combined acceleration in distance and direction method of Rosenbrock (1960). Herein one method of direct search is considered, namely that of acceleration in distance as derived by Hooke and Jeeves (1969).

Of the gradient search methods we will describe briefly the method of Davidon (1959) as it has been modified by Fletcher and Powell (1963), this technique appearing to be one of the most successful.

Hooke and Jeeves search

There are several variations of this method, and indeed a family of direct search techniques has been devised based on the original concept. This strategy of search involves the two tactical manoeuvres of an exploratory move and a pattern move. Referring to figure 3.8 and considering the simple case

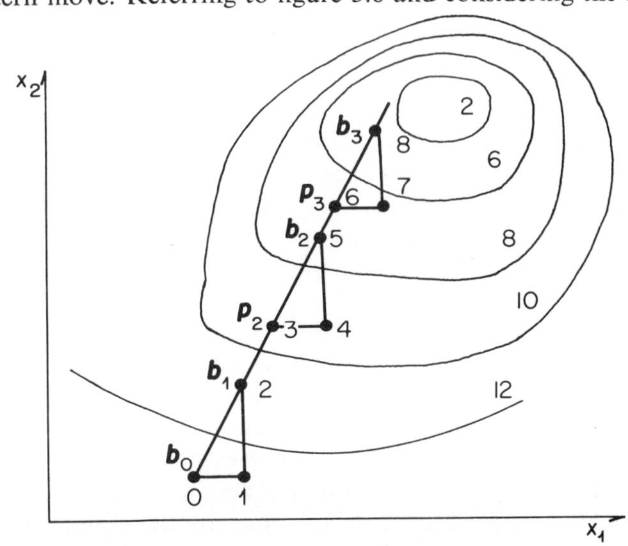

Figure 3.8 Direct search (Hooke and Jeeves, 1966).

of locating the minimum of a function of only two variables it is necessary to locate a series of 'base points' which are of pivotal importance. (Note that all points $\boldsymbol{b}_0$, $\boldsymbol{b}_1$ ... and $\boldsymbol{p}_1, \boldsymbol{p}_2$ ... are vectors representing the location of the points in n-dimensional space.)

The initial arbitrary base point is selected at $\boldsymbol{b}_0$. The next base point, $\boldsymbol{b}_1$, is located by the exploratory move from $\boldsymbol{b}_0$. Subsequently, all base points are found from a pattern move followed by an exploratory move. The exploratory move involves changing each variable in turn by a small increment testing to see after each change whether or not the situation has improved. (In this example this implies that an improvement is made if the function value is reduced since we are interested in locating the minimum of the function.) In figure 3.8 it can be seen that a small positive increment in x_1 is useful since the function value at station point 1 is lower than at station point 0 (here the station point term applies to any position at which the function is evaluated). Had this positive increment not proved to be useful then a negative increment would have been tried. If neither of these changes had resulted in any improvement then x_1 would have been kept the same and the next variable would have been similarly tested.

A positive increment added to x_2 is also beneficial, so the station point 2 becomes base point $\boldsymbol{b}_1$. (Note that a new base point is only located after all the variables have been tested.)

Base point $\boldsymbol{b}_2$ must now be located and this is found by a pattern move followed by an exploratory move. The pattern move is an extrapolation made according to the relation

$$\boldsymbol{p}_{i+1} = \boldsymbol{b}_i + (\boldsymbol{b}_i - \boldsymbol{b}_{i-1})$$

where $\boldsymbol{p}_{i+1}$ is an intermediate station point on the way to base point $\boldsymbol{b}_{i+1}$. Hence $\boldsymbol{p}_2$ in the diagram is located from

$$\boldsymbol{p}_2 = 2\boldsymbol{b}_1 - \boldsymbol{b}_0$$

In effect a vector step has been taken from the present base point equal to the vector step from the previous base point to the present one (because that step size proved beneficial we will try it again).

This pattern move is immediately followed by an exploratory move which may correct or extend the present pattern (the latter being so in figure 3.8). Hence a new base point $\boldsymbol{b}_3$ is found from the exploratory move starting at $\boldsymbol{p}_3$ ($\boldsymbol{b}_3$ corresponds to station point 8).

These selections permitted a successful approach to the optimum but if for example the exploratory search to base point $\boldsymbol{b}_3$ from $\boldsymbol{p}_3$ had been unproductive then $\boldsymbol{p}_3$ would be the new base point. After this, if it is obvious that the pattern move to $\boldsymbol{p}_3$ has been unsuccessful then the search returns to the previous base point, $\boldsymbol{b}_2$, where another exploratory move is made to start a new pattern of search. In the event of continued failure then the increment of variable change is reduced. Subsequently the search is designed to terminate when this increment size falls below some minimum value.

Davidon, Fletcher–Powell search

When considering gradient methods of search for finding the optimum of a function of several variables $\mathbf{x}_1$, $\mathbf{x}_2$ ..., $\mathbf{x}_n$ it is supposed that the function can be evaluated at all points and that the derivatives of the objective function with respect to each of the n variables, collectively known as the gradient of the objective function, can be calculated.

In order to show why a new approach to the problem of gradient search was required, consider the original gradient search methods available prior to the development of Davidon's method (1959). The basic gradient search method is that of deepest descent (assuming minimization of the objective function). This involves a cyclic search which approximates the continuous line of deepest descent to a series of straight lines. Each line corresponds to a stage in the search in which the local objective function gradient is determined and a set of moves is then made along the direction of steepest descent searching one-dimensionally for the approximate location of the optimum. This point serves as the base point for the next stage considering the next variable in a similar one-dimensional search.

In vector notation we can represent this procedure as

$$\boldsymbol{x}_i^{k+1} = \boldsymbol{x}_i^k - \lambda \boldsymbol{d}$$

where $\boldsymbol{x}_i^{k+1}$, $\boldsymbol{x}_i^k$ are successive points in the search considering variable x_i.
λ is a measure of the step length to the new base point.
$\boldsymbol{d}$ is a function of the partial derivatives with respect to each of the variables.

An alternative representation involves matrices,

$$\boldsymbol{X} = \boldsymbol{A} - \lambda \boldsymbol{I}\boldsymbol{D}$$

where

$$\boldsymbol{X} = \begin{bmatrix} x_1 \\ x_2 \\ \vdots \\ x_n \end{bmatrix} \quad \boldsymbol{A} = \begin{bmatrix} a \\ b \\ \vdots \\ n \end{bmatrix} \quad \boldsymbol{D} = \begin{bmatrix} \delta f/\delta x_1 \\ \delta f/\delta x_2 \\ \vdots \\ \delta f/\delta x_n \end{bmatrix}$$

n is an integer equal to the number of variables—a, b, ..., n, are equivalent to the base point values of x_1, x_2, ..., x_n for each step. I is the unit matrix which does not affect the equation but as will be seen its inclusion provides for conformity with the matrix equations derived for the other methods to be discussed. The method of steepest descent is extremely reliable but inherently slow in its approach to the minimum value of the objective function.

The method of Newton removes the disadvantage of a slow rate of convergence, but unfortunately this technique cannot be relied upon to reach the optimum. Using matrix notation Newton's method can be expressed as

$$\boldsymbol{X} = \boldsymbol{A} - \boldsymbol{J}^{-1}\boldsymbol{D}$$

where

$$J = \begin{bmatrix} \frac{\delta^2 f}{x_1^{\,2}} & \frac{\delta^2 f}{x_1 x_2} & \cdots & \frac{\delta^2 f}{x_1 x_n} \\ \vdots & & & \vdots \\ \frac{\delta^2 f}{x_n x_1} & \frac{\delta^2 f}{x_n x_2} & \cdots & \frac{\delta^2 f}{x_n^{\,2}} \end{bmatrix}$$

A further disadvantage of this approach is the necessity to compute the second derivatives of the function with respect to the independent variables.

Davidon succeeded in combining these two methods with the exclusion of the disadvantages of each, namely, the slow progress of the steepest descent method and the unreliability and the evaluation of second order derivatives of the Newton method. The matrix search equation became

$$\boldsymbol{X} = \boldsymbol{A} - \lambda \boldsymbol{H}\boldsymbol{D}$$

where the matrix $\boldsymbol{H}$ has an initial value of $\boldsymbol{I}$ (equivalent to the steepest descent analysis) and a final value, after n trials when the optimum has been located, of $\boldsymbol{J}^{-1}$ (the Newton method matrix).

The difficulty with this method is the way in which the matrix $\boldsymbol{H}$ is updated for each trial. One technique that is used is defined by

$$\boldsymbol{H}_{i+1} = \boldsymbol{H}_i + \frac{\boldsymbol{h}_i \boldsymbol{h}_i^T}{(\boldsymbol{h}_i^T \boldsymbol{y}_i)} - \frac{\boldsymbol{H}_i \boldsymbol{y}_i \boldsymbol{y}_i^T \boldsymbol{H}_i}{\boldsymbol{y}_i^T \boldsymbol{H}_i \boldsymbol{y}_i^T}$$

where

$$\boldsymbol{h}_i = \boldsymbol{X}_{i+1} - \boldsymbol{X}_i; \quad \boldsymbol{y}_i = \boldsymbol{D}_{i+1} - \boldsymbol{D}_i$$

The quadratic convergence that is applied near the optimum effectively means that

$$|\boldsymbol{h}_{i+1}| \propto |\boldsymbol{h}_i|^2$$

and not

$$|\boldsymbol{h}_{i+1}| \propto |\boldsymbol{h}_i|$$

Obviously convergence is more rapid in the former quadratic case. As a simple illustration of the Davidon–Fletcher–Powell technique consider the function which we wish to minimize as,

$$x_1^{\,2} - 2x_1 x_2 + 2x_2$$

and take $x_1 = -4$; $x_2 = 2$ as the initial base value of the variables, consequently $f(x_1, x_2) = 36$.

$$\boldsymbol{X}_0 = \begin{bmatrix} -4 \\ 2 \end{bmatrix} = \qquad \boldsymbol{D}_0 = \begin{bmatrix} -12 \\ 16 \end{bmatrix}$$

$$\boldsymbol{H}_0 = I = \begin{bmatrix} 1 & 0 \\ 0 & 1 \end{bmatrix}$$

$$\therefore \quad \boldsymbol{X}_1 = \begin{bmatrix} -4 \\ 2 \end{bmatrix} - \lambda \begin{bmatrix} 1 & 0 \\ 0 & 1 \end{bmatrix} \begin{bmatrix} -12 \\ 16 \end{bmatrix} = \begin{bmatrix} -4+12\lambda \\ 2-16\lambda \end{bmatrix}$$

a suitable value of λ is found by a linear search to minimize $F(\lambda)$, i.e. by setting $\frac{\mathrm{d}f}{\mathrm{d}\lambda} = 0$

where $F(\lambda) = f(X) = (-4+12\lambda)^2 - 2(-4+12\lambda)(2-16\lambda) + 2(2-16\lambda)^2$
hence $\lambda = 0.1923$

and $$X_1 = \begin{bmatrix} -1.692 \\ -1.077 \end{bmatrix}$$

and $f(x_1, x_2) = 1.5385$.

Now with $x_1 = -1.692$; $x_2 = -1.077$

$$D_1 = \begin{bmatrix} -1.231 \\ -0.923 \end{bmatrix}$$

and updating H then,

$$h_0 = X_1 - X_0 = \begin{bmatrix} 2.308 \\ -3.077 \end{bmatrix} y_0 = D_1 - D_0 = \begin{bmatrix} 10.769 \\ -16.923 \end{bmatrix}$$

and performing the necessary operations gives

$$H_1 = \begin{bmatrix} 0.781 & 0.361 \\ 0.361 & 0.412 \end{bmatrix}$$

Applying the linear search procedure again gives a new value of $\lambda = 1.309$. Hence

$$X_2 = \begin{bmatrix} -1.692 \\ -1.077 \end{bmatrix} - 1.309 \begin{bmatrix} 0.781 & 0.361 \\ 0.361 & 0.412 \end{bmatrix} \begin{bmatrix} -1.231 \\ -0.923 \end{bmatrix} = \begin{bmatrix} 0.0026 \\ 0.0022 \end{bmatrix}$$

and as can be seen since the function is quadratic in form the minimum (i.e. $f(X) = 0$) has been located after two cycles when both x_1 and x_2 are equal to zero (discrepancies in these values are due to rounding errors).

Fletcher and Powell (1963) illustrated the use of a powerful method with rapid convergence which was based on the original procedure described above as developed by Davidon. The authors in this particular reference consider the location of the local minimum of a multivariable function and show that convergence is always obtained most rapidly. There is also some discussion on the stability of the method and the concept of quadratic convergence as well as some explanation regarding the updating of the matrix H mentioned in the Davidon procedure. The conclusions arrived at confirm that this type of approach is considerably superior to other methods which are available.

This original article considered the objective function as unconstrained. However, Siddall (1970) has recently utilized this approach for constrained functions and has included a subroutine in the 'OPTISEP' package referenced in chapter 7. The search is concluded when the value of U defined as the objective function by

$$U = U(x_1, x_2 \ldots, x_n)$$

subject to

$$\phi_n(x_1, x_2, \ldots, x_n) > 0 \qquad k = 1, p$$
$$\psi_j(x_1, x_2, \ldots, x_n) = 0 \qquad i = 1, m$$

does not change significantly in two successive steps.

A more detailed discussion of the comparison and effectiveness of both direct search and gradient search techniques is given in Beveridge and Schecter (1970) which also indicates the scope and variety of numerical and analytical methods available for the optimization of univariable and multivariable functions both of restricted and unrestricted type.

3.4 Correlation of data

We will consider later the modelling of large systems such as the process and plant units. Lower down the hierarchy of systems there is an area of work in which it is important to correlate data by an equation. Of course such data may be left in graphical or tabular form but this is normally time consuming for the user or if stored in such fashion within the computer involves expense on storage files. Consequently considerable work has and is being carried out in order that previous reference sources make information available in equation form, e.g. cost of equipment, physical properties, material stresses. Similarly, new experiments are more readily correlated by equation fitting techniques, e.g. machinery performance trials, results of experimental work. As the latter is met by the student at an early stage of his training it is useful to give some guidance in this area, particularly as setting up appropriate programs to carry out such analysis is in itself a useful exercise.

Obviously for a set of data there is a boundless range of equation forms and the methods indicated may be used for the examples.

Simple equation fitting

To devise a polynomial equation such that

$$y = a_0 + a_1 x + a_2 x^2 \ldots + a_n x^n$$

where one variable, y, is a function of another variable, x, it is necessary to have available $(n+1)$ data points.

If the number of points exceeds $(n+1)$ it is usual to seek a 'best fit' of the data to a curve.

Consider the quadratic equation

$$y = a_0 + a_1 x + a_2 x^2 \qquad (n = 2)$$

Three data points $(y_1 x_1;\ y_2 x_2;\ y_3 x_3)$ permit the equations

$$y_1 = a_0 + a_1 x_1 + a_2 x_1{}^2$$
$$y_2 = a_0 + a_1 x_2 + a_2 x_2{}^2$$
$$y_3 = a_0 + a_1 x_3 + a_2 x_3{}^2$$

to be written and solution of these three (linear with respect to variable a) simultaneous equations give the values of a_0, a_1, a_2.

One of several different procedures involves the use of Lagrange interpolation. For instance the above quadratic can be written as

$$y = b_1(x-x_2)(x-x_3)+b_2(x-x_1)(x-x_3)+b_3(x-x_1)(x-x_2)$$

and setting $x = x_1$, x_2 and x_3 in turn gives

$$b_1 = y_1/((x_1-x_2)(x_1-x_3))$$
$$b_2 = y_2/((x_2-x_1)(x_2-x_3))$$
$$b_3 = y_3/((x_3-x_1)(x_3-x_2))$$

The general form for a polynomial of degree n is

$$y = \sum_{i=1}^{n+1} y_i \prod_{j=1}^{n+1} \frac{(x-x_j) \text{ omitting } (x-x_i)}{(x_i-x_j) \text{ omitting } (x_i-x_i)}$$

and to produce a flowsheet and program of this formula is recommended as an interesting project.

3.4.1 BEST-FIT METHOD

Where more than $(n+1)$ data points are available to fit to a polynomial of degree n a best-fit curve is drawn. One general procedure is the method of least squares in which the sum of the squares of the deviation is minimized. For the first order equation

$$y = a+bx$$

where m pairs of data points are available: $(x_1, y_1) \ldots, (x_m y_m)$, the deviation of any point from the values calculable by the equation is $(a+bx_i-y_i)$. The objective function is

$$\min \sum_{i=1}^{m} (a+bx_i-y_i)^2$$

by selection of the values a and b. Differentiating this function with respect to a and b and setting the derivative equal to zero gives

$$\Sigma 2(a+bx_i-y_i) \quad = 0$$

and

$$\Sigma 2(a+bx_i-y_i)x_i = 0$$

which may be rewritten as

$$\Sigma y_i \quad = ma+b\Sigma x_i$$
$$\Sigma x_i y_i = a\Sigma x_i+b\Sigma x_i^2$$

and simultaneous solution gives

$$a = \frac{\Sigma x_i^2 \Sigma y_i - \Sigma x_i \Sigma x_i y_i}{m\Sigma x_i^2 - (\Sigma x_i)^2}$$

and

$$b = \frac{m\Sigma x_i y_i - \Sigma x_i \Sigma y_i}{m\Sigma x_i^2 - (\Sigma x_i)^2}$$

Similar equations may be derived for higher order equations generally using the computer to solve the sets of simultaneous equations so developed (Stoecker, 1971).

3.4.2 FUNCTION OF MORE THAN ONE VARIABLE

Frequently a variable is a function of two other variables, such as motor revolutions and flow rate which can be used to determine the head developed by a centrifugal pump. These results are commonly presented as a family of curves for each pump speed. It may be desired to determine these as an overall equation. Thus a family of three curves may be represented by quadratic equations such that

$$y_1 = a_1 + b_1 x + c_1 x^2 \qquad \text{for variable } z = z_1$$
$$y_2 = a_2 + b_2 x + c_2 x^2 \qquad \text{for variable } z = z_2$$
$$y_3 = a_3 + b_3 x + c_3 x^2 \qquad \text{for variable } z = z_3$$

The relationship between the a constants may be expressed as a second degree equation in terms of z, i.e.

$$a = A_0 + A_1 z + A_2 z$$

using the data points (a_1, z_1), (a_2, z_2) and (a_3, z_3) and similarly using six more points

$$b = B_0 + B_1 z + B_2 z^2$$
$$c = C_0 + C_1 z + C_2 z^2$$

The general equation may be derived as

$$y = A_0 + A_1 z + A_2 z^2 + (B_0 + B_1 z + B_2 z^2)x + (C_0 + C_1 z + C_2 z^2)x^2$$

Note that for the general case the variables x, y, z may themselves consist of dimensionless groups of other variables such as may be derived by dimensional analysis. A typical example of such use is for correlating the performance of a compressor.

Similar considerations apply when analysing experimental results. Such data generally takes the form of a dependent variable which is influenced by several other variables often termed independent, say

$$v = f\ (x, y, z,)$$

The classical method of experimentation to separate the variables is to hold all known levels and determine the relation between say v and y at these conditions. However, if the separation of the variables is carried out in the mathematical analysis of the data greater freedom in experimental method is possible. Further unsuspected interactions can be obtained which lead to a better understanding of the processes (Brownlee, 1949).

A general equation may be selected to represent the physical law such as the expanded form of

$$v = \left\{ \begin{array}{l} \text{3rd order polynominal in } x \\ \text{2nd order polynominal in } y \\ \text{1st order polynomial in } z \end{array} \right\}$$

and multiple regression analysis similar to that in section 3.4.1 (A. M. Neville, J. B. Kennedy, 1964) may be used to check the significance of terms. Those found insignificant are eliminated and the data reanalysed.

The form of the equation should be predicted when using the method for curve fitting.

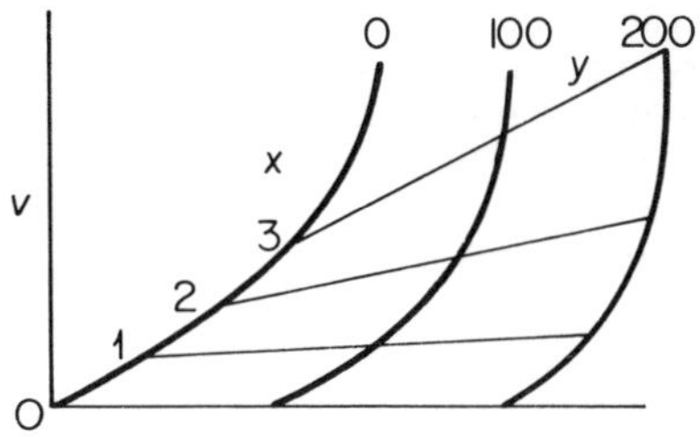

Figure 3.9 Plot of complex polynomial

Thus in figure 3.9, $v = 0$ for all values of y at $x = 0$, therefore no constant; y is linear; x is cubic. Expanding v as a 3rd order polynomial in x, 1st order polynomial in y, gives

$$v = \Theta_1 x + \Theta_2 x^2 + \Theta_3 x^3 + \Theta_4 xy + \Theta_5 x^2 y + \Theta_6 x^3 y$$

which may then be correlated to determine constants.

Higher order polynomials should be avoided in correlations as they usually oscillate. Polynomials go to + or − infinity as x is large. If the curve asymptotically approaches a constant at large values of x such a curve (as in figure 3.9) cannot be fitted. An attempt should then be made to try fitting the polynomial to the reciprocal $1/x$ or an exponential.

Arbitrary functions

If the function cannot readily be obtained as an analytical function it may be stored as an arbitrary function. The coordinates of points on the curve are established by a series of readings as deemed necessary by the shape of the curve and stored in array form in the program. Consider an arbitrary function relating Y to X. Let the data be stored in some form

DATA (X(N),N = 1,10)/0.,0.5,1.,2.,6.,6.5,7.,8.,9.,15./

where 10 points are read for the corresponding values of Y

DATA (Y(N),N = 1,10)/1.,3.,5.,7.,9.,11.,13.,15.,18.,20./

If the value of Y is required at some value of X in between those stored above it may be calculated by interpolation between the points which surround this particular value. Thus

$$y = y_{i-i} + \frac{(x - x_i)}{(x_i - x_{i-1})}(y_i - y_{i-1})$$

where $i = 2,N$.

The value is obtained by writing a function subroutine which evaluates the function in this manner. The method can readily be extended to functions of more than one variable.

Notation	Definition	Section
a_i	polynomial coefficient	general
$a_{i,j}$	linear equations coefficients	3.2
a, b	search region boundary limits	3.3.1
a, b, c	quadratic coefficients	3.4
adj	adjoint matrix	3.2
$A_{i,j}$	cofactor matrix element	3.2
$\boldsymbol{A}$	matrix	3.2
A, B, C	quadratic coefficients	3.4
$\boldsymbol{b}$	Hooke and Jeeves vector	3.3.3
b	Wegstein promotor variable	3.1.5
C_n	linear equation constant	3.2
$\boldsymbol{d}$	partial derivatives vector	3.3.3
e	trial point increment	3.1
f	function	general
$\boldsymbol{g}$	constraint limitations vector	3.3
I_i	interval of search	3.3.1
k_s	convergence promotor enhancement factor	3.1.5
l	search point distance from region boundary	3.3.1
L_K	length of region of interest	3.3.1
m	number of pairs of data points	3.4.1
n	number of points	3.4
$\boldsymbol{p}$	Hooke and Jeeves vector	3.3.3
$\boldsymbol{p}$	parameter vector	3.3
$\boldsymbol{q}$	decision variable vector	3.3
q	Wegstein promotor variable	3.1.5
r_i	root value	3.1
U	objective function	3.3.3
V	variable	3.3.2
x	variable	general
y	variable	general
z	variable	3.4.2

Greek symbols

Notation	Definition	Section
$\alpha_{i,j}$	inverse matrix element	3.2.1
δ	partial differentiator	general
ϕ	objective function	3.3
ϕ_n	inequality constraint function	3.3.3
ψ_j	equality constraint function	3.3.3
λ	Davidson–Fletcher–Powell step length	3.3.3
Θ_i	polynomial coefficient	3.4.2

Subscripts

Notation	Definition	Section
i	row number in a matrix or iteration step	general
j	column number in a matrix	general
r	root value of variable	general

Superscripts

n	polynomial degree	general
$'$	first derivative	general
$''$	second derivative	general
r	trial point	general
$*$	best value	general

4 Developing a Unit Computation

In order to show the development of a program one example is considered, a category B program on distillation. The further development of the design toward that of the category C program is indicated.

4.1 Program organization

Any program should be written to a given standard which will permit its use by personnel other than the programmer. The use of comment statements throughout the program is desirable as, too, is a description of the general technique. However, it is the peripheral features which determine the program's acceptance and general application. These are aspects such as input, output, monitoring and ability to process several sets of data in sequence.

4.1.1 DATA INPUT

The formatting requirement must be carefully defined. For interactive programs some form of question and answer system may be used. For batch jobs the system is usually tabular and clear instructions as to the sequence of providing information are required. A standard format for all data may be an advantage even if it seems ludicrous to express the number 1.0 in say F5.2 FORMAT. This is, of course, not always possible. Different people's conception of what is 'easy to use' varies widely and instead the programmer should try to make the program difficult to mis-use.

The sequence in which data is prepared depends on individual programs, but we have found that grouping as follows helps the inexperienced to avoid many errors.

Job description: self-obvious.
Number of runs: change card to permit varying parameters from base case.
Internal program control: calling sequence for subroutines, monitoring, output format.
Process specifications: input streams, stream specifications.
Unit parameters: geometric, constructional and restrictive specifications.
Physical properties: direct input or access to physical properties package.

The greatest problem in data input is not in a user failing to obtain satisfactory results on his first run through the program, but that errors enter into calculations and are not discovered. Thus a data check system should be used which covers dimensional discrepancies or omissions interpreted as intentional zeros.

As the program passes through strategic nodal points or iterative loops, provision should be made for print-out, either for debugging, a check on whether the result is consistent, or a warning message. The latter may be programmed to be printed out when the program fails in which case the user may take remedial action. Alternatively, despite the warning a terminal solution may be permitted, in which case the user should analyse in this area to ensure against errors in logic.

For investigation of different values of design variables a form of change system is required. This can be designed so that only key parameters are altered and input of a complete set of data for each case studied is avoided. The program may be directed to a stage part way into the program at which this data is used.

4.1.2 DATA OUTPUT

The user should be able to control the extent of data output. In the early stages of debugging extensive information may be required but at later stages or even to other users excessive diagnostic output can cause considerable confusion.

The input data should be reprinted. Summary output should be in a form which can be used readily by the engineer. This may be in a tabular form showing the effect of design variables for a series of cases given for manual optimization. A preferred form for equipment is as a specification sheet for transmission to fabricators, etc.

4.1.3 PROGRAM ASSEMBLY

The above principles and the methods by which experience can be obtained in program assembly are indicated using a distillation unit as an example.

The first contacts of a student with distillation as a separation process is usually restricted to binary ideal systems, which give an introduction to the concepts of equilibrium and relative volatility. The solutions to these problems are normally presented as analytical or graphical, this being possible since the design equations required are well defined and sufficient significant variables are known, so that no initial estimates or interactive calculations are necessary to arrive at the solution.

In order to progress from this position to the more realistic, but complex, multicomponent, non-ideal systems it is necessary to have a complete understanding of the fundamental heat and mass balances and equilibrium considerations introduced with the binary systems. Boundaries must be defined for the solution since the computer in the logical search for a solution must be aware of the physical limitations imposed on the system because of the properties of the components and the feasibility of operation of the unit. In general terms the human element of decision has been removed from the problem and it is imperative that these boundaries should be defined to avoid divergent calculations or impractical conditions being generated.

4.2 A case study: sizing a distillation column

4.2.1 GENERAL

As an example of the procedure for generating a computer program for a distillation unit first consider the Fenske–Underwood–Gilliland method which is used to provide an initial size estimate of the unit. Later using a similar approach a more complex program is developed for the design of the unit based on the rating method of Thiele–Geddes (1933). The Fenske program is listed in the appendix. The calculation is carried out in the sequence shown in figure 4.1 and will be discussed in the sections indicated in this diagram. Correct order of solution is important and the use of flowcharts to outline the method of solution is essential. It is necessary to isolate the system variables already defined for the unit. Consider the physical state of the feed,

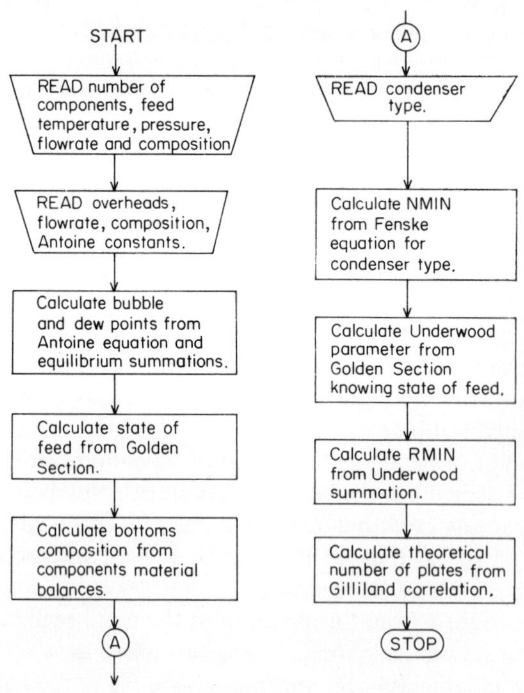

Figure 4.1 Overall flowchart (Fenske–Underwood–Gilliland).

i.e. temperature, pressure and composition; and the flowrates of overhead product required including the fraction of key components to be recovered. This latter factor is particularly important since it defines, with the reflux ratio, the ultimate size of the column. It is necessary to find the fraction of the feed that flashes on entry into the column if not previously defined, this parameter being required for the Underwood summation (1945) (the state of the feed or q-value). A check is made to find whether or not the feed temperature, at the defined conditions, lies within the range of bubble dew points. If this temperature is less than the bubble point or greater than the dew point then the fraction of feed that is vapour will be assumed to be either 0.0 or 1.0 respectively. This approximation is to some extent justifiable since the sensible heat contribution of either subcooled liquid or superheated vapour will be small in comparison with the latent heat of the feed.

4.2.2 BUBBLE AND DEW POINTS

The distribution coefficient is calculated, assuming an ideal system, from the Antoine equation for vapour pressure of a pure component:

$$K = \frac{P^*}{P} = \exp((A-B)/(T+C))^*\ 2.303)/P$$

where K = distribution coefficient
P = total pressure mm Hg
P^* = vapour pressure of pure component mm Hg
T = temperature °C
A, B, C = Antoine constants

The Antoine constants must be supplied as input data (the alternative being to have physical data for components available in data banks). The assumption of the ideal system, Raoult's law, can be avoided if fugacity coefficients are included in the above equation but for many systems, especially hydrocarbon series, this assumption is reasonably valid.

The bubble point temperature is the value for which,

$$\sum_{i=1}^{M} K_{iF} z_i = 1$$

and similarly the dew point is found from,

$$\sum_{i=1}^{M} \frac{z_{Fi}}{K_i} = 1$$

where z_{Fi} = total mole fraction of component i in feed
K_i = distribution coefficient for component i
M = number of components

The convergence calculation involved in these summations may best be explained by reference to flowchart, figure 4.2. The principal of the convergence being the evaluation of the respective equilibrium summation expressions for two temperatures T and $T+0.001$. Depending on the difference in values of these two trials a new T value is calculated according to the Secant method. This generates a new value depending on a weighted error correction term, i.e.

$$T = T+(0.001*(1-SX1)/(SX2-SX1))$$

When this new value differs from the first estimate value of the previous iteration by a margin defined by the user (here less than 0.001), then the calcula-

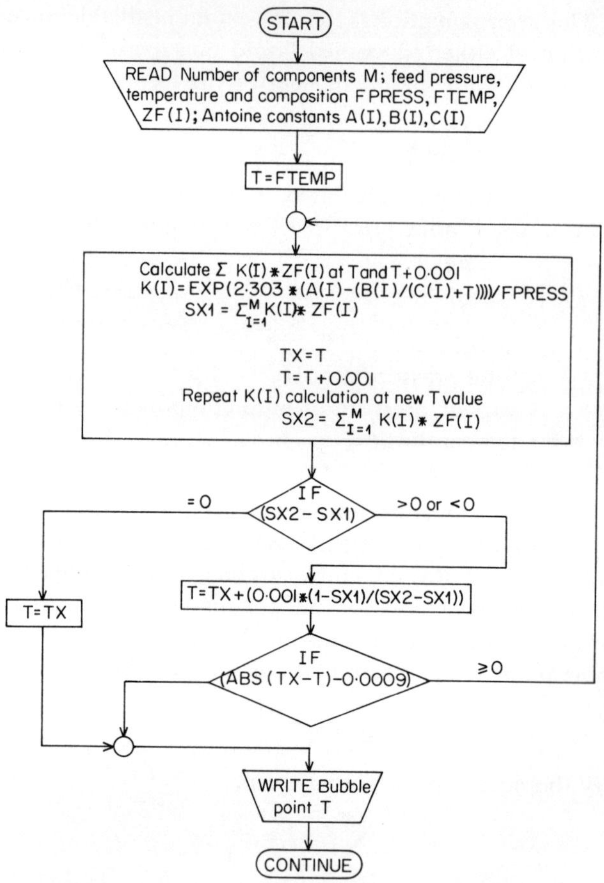

Figure 4.2 Flowchart for bubble point calculation.

tion is considered to have converged to the required temperature. The flowchart illustrated is for the bubble point calculation, the dew point case being the same except for the form of the summation. In both cases an initial temperature estimate is required to start the calculation and for convenience this is taken to be the feed temperature.

As an alternative the Newton–Raphson technique may be used to promote the new temperature using

$$T = T + \left(1.0 - \sum_{i=1}^{M} \frac{z_{Fi}}{K_i}\right) \Big/ \left(\sum_{i=1}^{M} z_{Fi} * B_i/(T + C_i)^2\right)$$

where B, C are the Antoine constants for component i.
This expression is valid for the dew point determination and the bubble point can be found using a similar expression after redefining the summation term.

Unfortunately, as is inherent with this approach, divergence has been found to occur for particular systems whereas the Secant method has generally produced convergence to the required temperature values.

Having established the limits of temperature, the fraction of feed that is vaporized must be found for the Underwood q-value using the assumptions indicated earlier.

4.2.3 VAPOUR FRACTION OF FEED

To show alternative procedures the Golden Section method of logical search is used for this calculation since the function which will be tested is unimodal (has only one minimum value), a necessary condition for this search method to be valid (3.3.1).

The function that is to be tested is,

$$\text{Absolute}\left(\sum_{i=1}^{M} \frac{z_{Fi}}{1 + \dfrac{L}{VK_i}} - V\right)$$

where V = fraction of feed that is vapour
L = fraction of feed that is liquid $(1 - V)$

This function can be defined from a simple component balance around the feed position. Inclusion of the absolute extended function means that the problem of negative values is eliminated so that the minimum value sought is in fact zero. The flowchart for this calculation is shown in figure 4.3. On the flowchart E and F represent the boundaries of the search region, respectively 0.0 and 1.0. The variables G and H are the actual values of vapour fraction that are tested in the function defined. The corresponding function values, EE and FF, are found, and depending on the sign of the difference between these values a section is eliminated from the feasible region of search and the calculation is repeated. This comparison and elimination continues until the values G and H are within the user defined acceptable error margin (here 0.0005) whereupon the calculations are terminated.

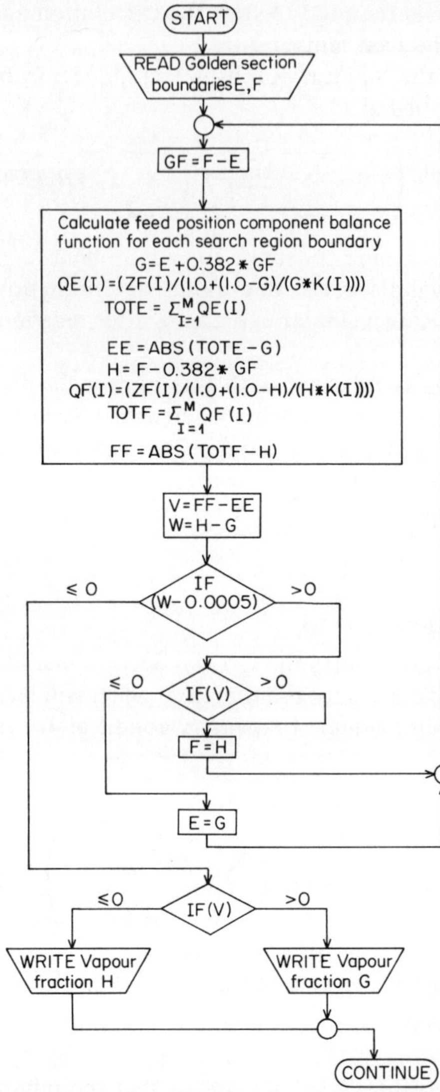

Figure 4.3 Flowchart for vapour fraction of feed calculation.

4.2.4 FENSKE—MINIMUM NUMBER OF PLATES

The minimum number of theoretical plates can be found from the Fenske equation for a total condenser (1932),

$$N_{MIN} = \log \frac{\left[\left(\frac{x_{T1}}{x_{T2}}\right)\left(\frac{x_{B2}}{x_{B1}}\right)\right]}{\log \alpha_1} - 1$$

where x_T = overheads composition mole fraction
x_B = bottoms composition mole fraction
α = relative volatility.

Subscripts 1 and 2 refer respectively to the light and heavy keys.

Note that the identification of the key components in any multicomponent system is of prime importance since often the requirements of the unit will be that a given percentage of the keys must be recovered in each product or that a certain split of keys is obtained. In a multicomponent system when lighter than light key and heavier than heavy key components are present, the keys can be identified within the computer program by assigning them particular locations in the variable matrices to the keys, i.e. designate particular subscript numbers to the keys and ensure that any physical property data supplied to the program lists the components to accommodate this restriction. The bottom key's composition and the component relative volatilities can be calculated within the program from mass balances and distribution coefficient ratios respectively. The relative volatilities being defined by

$$\alpha_i = \frac{K_i}{K_{ref}}$$

where K_{ref} is the reference component distribution coefficient which in this system is taken to be the heavy key, and calculated at an average temperature (again for convenience the feed temperature is used). Although only the light key volatility is required for the Fenske equation, the other values are used in the Underwood summation. The equations are programmed in logical order.

4.2.5 UNDERWOOD—MINIMUM REFLUX RATIO

The next stage of calculation involves the evaluation of the Underwood parameter, Θ. This problem is again suited to solution by use of the Golden Section search method, a similar type of function as for the vapour fraction calculation being derived. Rearrangement of the summation produces

$$\text{Absolute}\left(\sum_{i=1}^{M} \frac{\alpha_i z_{Fi}}{\alpha_i - \Theta} - (1-q)\right)$$

where Θ is the Underwood parameter
q is the state of the feed.

The search is for the minimum value of the above function (i.e. zero) where Θ is the variable under examination. The boundaries of the search region are defined, from the theory of Underwood, as being the relative volatilities of the light and the heavy key components (this implies that the lower limit is 1.0—the heavy key relative volatility). As before the search is concluded when the two values of Θ under examination are within an acceptable tolerance. Obviously the lower the tolerance the more iterations required to achieve the converged solution. Further details of this example are given in section 7.2. With the value of Θ calculated the minimum reflux ratio is found from the second Underwood summation,

$$R_{MIN}+1 = \sum_{i=1}^{M} \frac{\alpha_i x_{Ti}}{\alpha_i - \theta}$$

4.2.6 GILLILAND—ACTUAL NUMBER OF PLATES

The actual reflux ratio can then be defined as R_{MIN} multiplied by some factor, usually in the range of 1.1 to 1.8. The number of theoretical plates is found from the Gilliland (1940) correlation which is expressed in graphical form where $(N-N_{MIN})/(N+1)$ is plotted against $(R-R_{MIN})/(R+1)$. This 'curve' can best be accommodated into the computer program by dividing the graph into distinct regions (defined by particular values of the ordinate) and deriving a separate linear equation for each region which approximates to the curve, the smaller the regions then the more accurate the approximation becomes. An example of the input/output for this program as described is given in table 4.1. In order to produce a sizing of the column an efficiency term can be included to generate the actual number of plates which leads to a column height, and an empirical equation such as that of Souders and Brown (1934) yields a vapour velocity and hence a column diameter.

Further developments of the program could include, for example, feed plate location, the use of a partial condenser and the effect on cost of using a different factor in the reflux ratio value.

4.2.7 TRAINING IN PROGRAM ASSEMBLY

This elementary example indicates how a computer program is produced. *It is obviously essential that the programmer should have a complete understanding of the problem before the program is written.* With this knowledge ordering of the solution is relatively easy. Each particular part of the problem can be written as an individual section within the program, including where necessary numerical search methods and convergence techniques, and in this way the program is logically constructed. This concept of logical ordering is of prime importance especially when people, other than the writer of the program, may have reason to be involved in its use. For this reason, repeating an earlier suggestion, it is usually of value to employ comment statements throughout the program indicating what each section is concerned with, and even the programmer himself may find this to be a blessing if he returns to the program sometime in the future.

This example has been explained in sections representing the calculations which are combined to produce one overall program; this is a normal

INPUT DATA:
Number of components = 7
Feed temperature = 130 °C
Feed pressure = 1 276.72 mm Hg
Feed flowrate = 9 16.58 kg mol/h
Overheads flowrate = 277.49 kg mol/h

Component numbers	Total feed composition	Overheads composition	Antoine constants *A*	*B*	*C*
1	0.3017	0.9997	6.9057	−1211.033	220.790
2	0.3105	0.0003	6.9546	−1344.800	219.462
3	0.0761	0.0	6.9577	−1424.255	213.206
4	0.0548	0.0	6.9905	−1453.430	215.307
5	0.1328	0.0	7.0091	−1462.266	215.105
6	0.0656	0.0	6.9990	−1474.679	213.686
7	0.0575	0.0	7.0438	−1573.267	208.564

OUTPUT: after each stage of calculation
CASE STUDY BENZENE/TOLUENE/XYLENES, ETC
NUMBER OF COMPONENTS = 7
BUBBLE POINT OF FEED = 124.3555 DEGREES C
DEW POINT OF FEED = 144.1062 DEGREES C
FRACTION OF FEED THAT IS VAPOUR = 0.2999
MINIMUM NUMBER OF PLATES = 20.473
UNDERWOOD PARAMETER = 1.5555
MINIMUM REFLUX RATIO = 2.3264
FOR REFLUX RATIO OF RMIN * 1.5 THAT IS = 3.4896
NUMBER OF PLATES = 35
FEED PLATE NUMBER IS = 22

Table 4.1 Input and output of Fenske–Underwood–Gilliland.

practice of program building. An alternative procedure is to build the program as a series of subroutines, as say for the bubble point calculation which in more rigorous plate to plate programs can be called for the conditions pertaining on every plate. An advantage of this method is that this system of subroutine assembly permits a more efficient method of debugging before evaluating the complete program. When the program has been successfully compiled (i.e. programming errors removed) a TRACE system can be included to locate execution errors due to incorrectly defined data. The use of subroutines in a large program means that alternative methods of solution can easily be substituted by replacement of the relevant subroutines without affecting the complete program, e.g. replacing the Golden Section search method by a dichotomous search.

Calculation segments such as that of bubble point estimation can easily be adapted as subroutines and the modification of the above example into a system of a calling segment and individual section subroutines is suggested as a useful exercise.

A further approach for gaining expertise in handling large subroutine programs is to introduce the concept of 'black boxes' each of which represents a subroutine. In this way programs can be built from these units provided that clear comment documentation is shown to indicate the input supplied and the output required for each subroutine.

4.3 A more complex method—Thiele–Geddes solution

4.3.1 INTRODUCTION

Without entering into an explanation of the theory involved in this rating method of distillation design (there are many references to this in the classical equilibrium stage texts), it is necessary to define certain parameters of the system before the analysis can be attempted. The number of trays, the feed plate location, the feed analysis and the reflux ratio must be defined along with estimates for the overheads and bottoms compositions. These estimates are required to produce an initial starting point for the calculations. In this way a detailed plate to plate analysis of the unit is generated, i.e. liquid flow, vapour flow, temperature and composition profiles.

The previously defined method of Fenske–Underwood–Gilliland is an ideal subroutine to generate the initial input variables required for the more complex program and the authors have recently been developing a program along these lines, using the rating method essentially as a check to validity of the short-cut method of design. When the compositions of the products differ from the initial estimates then further stripping or rectifying plates are needed to effect the required split.

4.3.2 PROGRAM DESCRIPTION

With the availability of digital computers it is less of a necessity to consider the quantity of calculation that is required. This means that the mathematical model chosen to describe the system can be comprehensive enough to include both complex and conventional columns. The particular program under discussion allows for stages within the column as well as the reboiler and condenser (either partial or total). Each plate within the column can have a feed inlet, a sidestream outlet and provision for either heat supply or removal.

The design of a distillation column on a plate-to-plate analysis depends on the ability to converge the respective flow rates, temperature and composition profiles according to the balances and equilibrium restraints. The fundamental balances of overall material, component material and energy as defined for the simple binary system are equally applicable in the multicomponent case, but a considerable amount of iterative calculation is necessary before all these relationships are satisfied for each plate within the column.

A diagram of the system is shown in figure 4.4. The corresponding balances are:

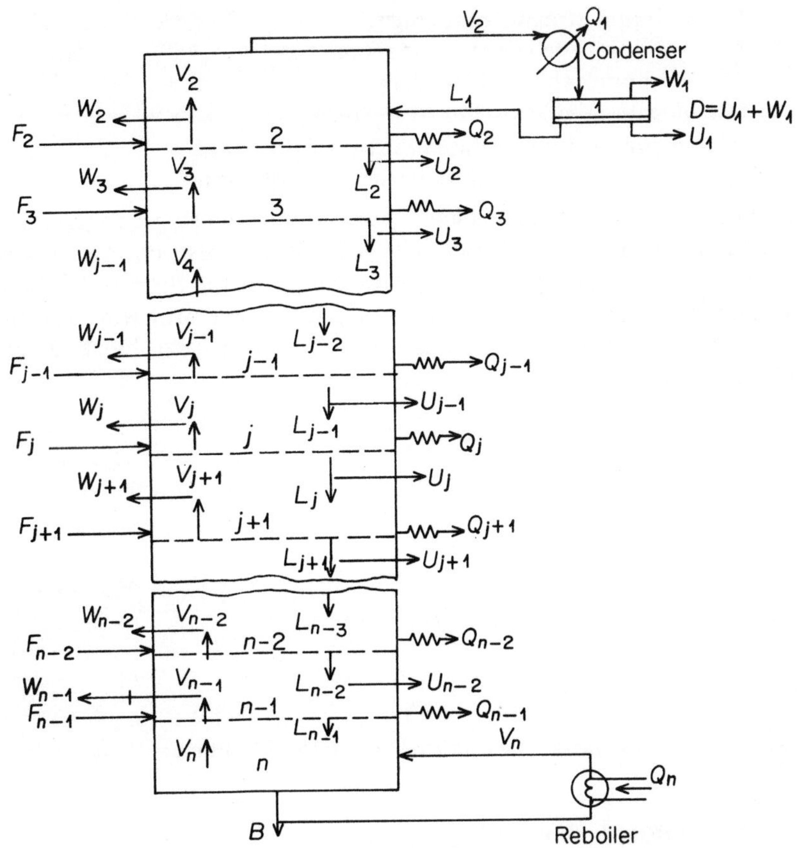

Figure 4.4 Flow nomenclature in a distillation column.

Overall material balance

$$(F_j + V_{j+1} + L_{j-1}) = (V_j + W_j + L_j + U_j)$$

for $2 \leqslant j \leqslant n-1$

Component material balance

$$(F_j z_{i,j} + V_{j+1} y_{i,j+1} + L_{j-1} x_{i,j-1}) = (W_j + V_j) y_{i,j} + (L_j + U_j) x_{i,j}$$

for $1 \leqslant i \leqslant m$ and $2 \leqslant j \leqslant n-1$

Energy balance

$$(F_j H_{Fj} + V_{j+1} H_{j+1} + L_{j-1} h_{j-1}) - [(W_j + V_j) H_j + (L_j + U_j) h_j) = Q_j$$

$2 \leqslant j \leqslant n-1$

where F_j = feed stream at stage j
H_{Fj} = enthalpy of feed stream F_j
H_j = enthalpy of vapour stream from stage j
h_j = enthalpy of liquid stream from stage j

L_j = liquid stream from stage j
U_j = liquid side stream from stage j
V_j = vapour stream from stage j
W_j = vapour side stream from stage j
$x_{i,j}$ = composition of liquid stream from stage j
$y_{i,j}$ = composition of vapour stream from stage j.

Specific balances can also be defined for the extreme stages in the system, namely the reboiler and condenser. The only further restriction with which to comply is that of the equilibrium relationships, in the form of the bubble point summation previously defined in the short-cut program. The flow-chart for the program is represented by the generalized block diagram flowchart, figure 4.5.

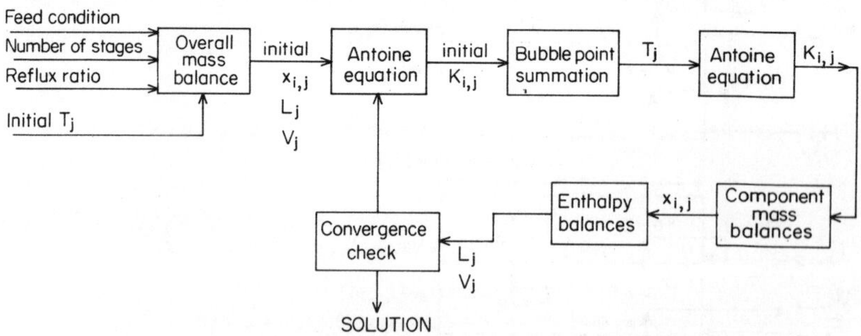

Figure 4.5 Generalized flowchart of column.

4.3.3 COMPUTATIONAL PROCEDURE

The iterative plate-to-plate procedure adopted for the solution of the balances and summations is illustrated in figure 4.6 and defined below:

(1) Calculate feed condition, reflux ratio and number of stages from subroutine FENSKE.
(2) Compute bottoms composition from overall mass balance—hence estimate initial $x_{i,j}$ profile (assumed linear) and initial L_j and V_j profiles (assumed constant in each of two sections of column).
(3) Calculate initial $K_{i,j}$ values for every plate assuming all temperatures equal to the feed temperature.
(4) Using Secant method find temperature on every plate from bubble point summation.
(5) Calculate true $K_{i,j}$ values for all components on each plate.
(6) Compute new $x_{i,j}$.
(7) Normalize $x_{i,j}$.
(8) Determine new L_j and V_j values from enthalpy balances.
(9) Have all the values of V_j converged? If so, then calculation is complete. If not, then reset L_j and V_j equal to values from 8 and return to 3.

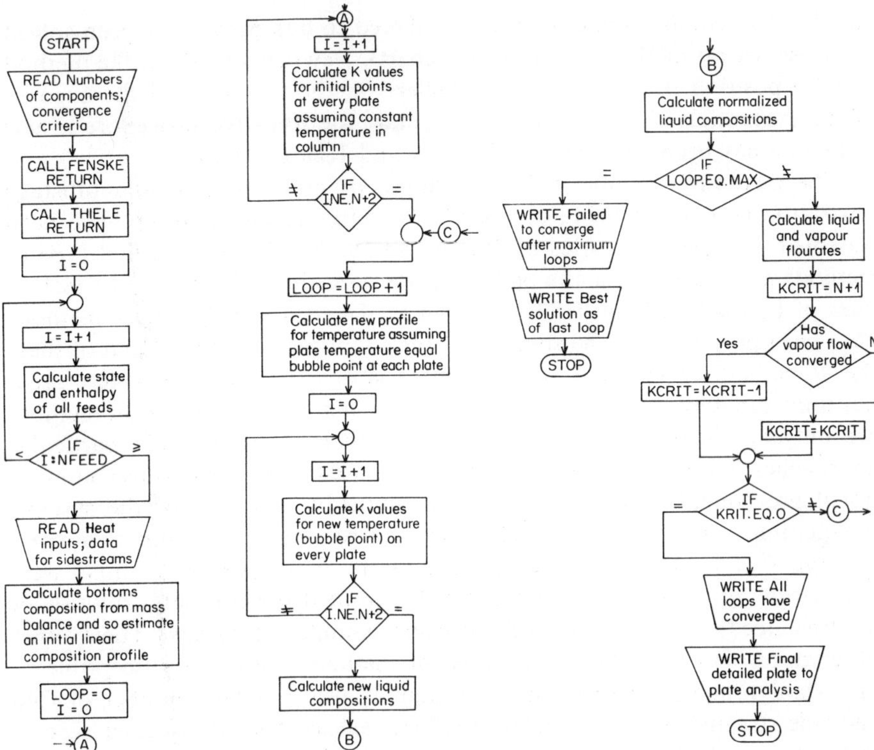

Figure 4.6 Flowchart: Thiele–Geddes column design.

4.3.4 IMPROVED METHODS OF ANALYSIS

Work has been carried out in this field to reduce the cost of computation and it is useful to study the method of attack.

An alternative method of analysis devised originally by Amundsen, Pontinen and Tierney (1959), greatly reduces the number of iterations, and consequently the amount of computing time, by describing the problem as the solution of a series of non-linear simultaneous equations. This idea was further developed by Wang and Henke (1966) with the introduction of the tridiagonal matrix algorithm and Muller's method (1956). It would appear that this development is more favourable and the equations required for the formation of the matrix can be divided into three groups:

(1) Component material balance equations.
(2) Energy balance equations.
(3) Summation equations.

As with all the methods of iterative convergence the solution of the model depends upon finding the values of T_j, L_j and V_j such that all the material and energy balances are satisfied. The problem of finding T_j so that the bubble point equilibrium condition is adhered to has previously been solved using the Secant method (as in the flowchart), Newton's method, the method of false position or more recently Muller's method. To avoid the expense of the evaluation of three functions to start the iterative procedure a new technique has been described by Houtby and Khan (1972).

The methods available for the solution of multicomponent multistage separation equipment using computerized rigorous calculations have been classified into two categories by Wang and Henke (1966). The first type of methods are obtained for the plate-to-plate procedures based on the techniques of Lewis–Matheson (1932) and Thiele–Geddes (1933). Many methods have been proposed for the solution of the equations so defined and of these the 'θ-method' of Holland and coauthors is one of the most widely used, but unfortunately this and other techniques, such as the relaxation method, are only applicable to problems that lend themselves to these particular approaches. The second type of methods are based on new rigorous computational procedures developed for the simultaneous solution of the systems of equations. Friday and Smith (1964) considered the relative merits of these two types of analysis and concluded that the first type of methods were numerically unstable and as such basically unsuited for computer solution of multiple feeds and multiple side streams complex columns. Their other observation was that the second type of methods were not susceptible to important round-off errors and work equally well for any number of feed and side streams. Wang and Henke (1966) realized that the second type of methods, which were then in existence, were either too slow or required a considerable amount of computer storage and furthermore that most of them used the Newton–Raphson method for convergence and unfortunately there is no guarantee of convergence. Consequently they developed the tridiagonal matrix method for the simultaneous solution of the linear component balance, energy balance and summation equations.

The tridiagonal matrix method involves the solution of an augmented matrix by the Gaussian elimination method (as described in section 3.2). The augmented matrix is of the form,

$$\begin{bmatrix} B_1 & C_1 & & & & D_1 \\ A_2 & B_2 & C_2 & & & D_2 \\ & A_j & B_j & C_j & & D_j \\ & & A_{n-1} & B_{n-1} & C_{n-1} & D_{n-1} \\ & & & A_n & B_n & D_n \end{bmatrix} \qquad 1 \leqslant i \leqslant m$$

this is obtained by combining the tridiagonal matrix $\boldsymbol{A_{BC}}$ with the column vector $\boldsymbol{D}$

$$B_1 = -(V_1 K_{i1} + U_1); \quad C_1 = V_2 K_{i2}; \quad D_1 = 0$$

$$A_j = L_{j-1} = V_j + \sum_{k=2}^{j-1} (F_k - W_k - U_k) - D$$

$2 \leqslant j \leqslant n-1$

and

$$
\begin{aligned}
D_1 &= B_1 x_{i1} + C_1 x_{i2} \\
D_j &= A_j x_{i,j-1} + B_j x_{ij} + C_j x_{i,j+1} \qquad 2 \leqslant j \leqslant n-1 \\
D_n &= A_n x_{i,n-1} + B_n x_{in}
\end{aligned}
$$

(the variables are as defined for figure 4.4, A_j, B_j, C_j and D_j being derived from the combination of the energy and material balances and equilibrium relationships). The problem can be defined in matrix equation form as the solution of

$$
\begin{bmatrix}
B_1 & C_1 & & & \\
A_2 & B_2 & C_2 & & \\
 & A_j & B_j & C_j & \\
 & & A_{n-1} & B_{n-1} & C_{n-1} \\
 & & & A_n & B_n
\end{bmatrix}
\begin{bmatrix}
x_{i1} \\ x_{i2} \\ x_{ij} \\ x_{i,n-1} \\ x_{in}
\end{bmatrix}
=
\begin{bmatrix}
D_1 \\ D_2 \\ D_j \\ D_{n-1} \\ D_n
\end{bmatrix}
$$

for each component ($1 \leqslant i \leqslant m$).

When the flowrates and compositions of all feed streams and the amounts of product stream are known then an initial set of V_j and T_j is assumed. Solution of x_{ij} is then obtained by the elimination method mentioned. With the values of $K_{i,j}$ (expressed as a function of T_j, as, for example, by the Antoine equation) the summation equation can be computed and new T_j's calculated by the method of Muller, mentioned previously. The new T_j values are used to compute a new set of V_j values from the energy balance equation before the whole process is repeated to find a new set of $x_{i,j}$ values. Convergence is obtained when the T_j sets of two successive iterations differ only by some predescribed tolerance.

The main advantage of this method over the general matrix method is that it is not necessary to invert the huge matrix in the solution of the matrix equation for the set of $x_{i,j}$ values. One further advantage is the greater certainty that the method of Muller will provide for convergence compared to techniques such as that of Newton–Raphson where there is no assurance of convergence, and with this latter method there is also the problem of evaluation of partial derivatives.

Unfortunately, this method of Wang and Henke incorporating the Muller technique cannot be used for absorbers and extractors (as explained by Friday and Smith). Consequently Stainthorp and Whitehouse (1967) proposed a procedure which consists of solving the material and energy balances as non-linear simultaneous equations. This approach has been refined by other workers and it can be said to be applicable to absorption and extraction as well as distillation. However, the method involves large computer capacity and has been found to be less efficient for distillation problems as compared with the conventional iterative approach. Overall, considering the relative merits of these methods suggests that the iterative approach of Wang and Henke is to be favoured especially for distillation applications.

Houtby and Khan describe how the original method of Wang and Henke can be made to be more efficient by:

(1) Reducing the number of column iterations. This is achieved by using a numerical acceleration technique to readjust the stage temperatures.

(2) Reducing the time per iteration. This is achieved by introducing a new numerical technique which instead of requiring three function evaluations (as in Muller's method), needs only two before commencing the iterative cycle. Furthermore this technique does not require the evaluation of any derivatives and thus permits the use of complex equilibrium correlations of any kind. This method has been applied to a wide cross-section of column configurations covering all the usual combinations of multiple feeds, side streams and condenser variations. Equilibrium relationships for multicomponent systems have been used in the form of polynomials, Antoine constants, Wilson coefficients and Chao-Seader correlations. The savings arising from use of this method become obvious when for example a large executive-module flowsheeting program, such as described in chapter 6, is to be processed and several column designs and configurations must be investigated.

Notation	Definition	Section
A, B, C	Antoine constants	4.2.2
A, B, C	tridiagonal matrix elements	4.3.4
B	bottoms product flowrate	general
D	overhead product flowrate	general
D	column vector element	4.3.4
F	feed flowrate	general
H_F	enthalpy of feed stream	4.3.2
H	enthalpy of vapour stream	4.3.2
h	enthalpy of liquid stream	4.3.2
K	distribution coefficient	general
L	fraction of feed that is liquid	4.2.3
L	liquid flowrate	4.3.2
M, m	number of components	general
N, n	number of plates/stages	general
P	pressure	general
P^*	vapour pressure of pure component	4.2.2
Q	heat duty	4.3.2
q	state of feed	general
R	reflux ratio	general
T	temperature	general
U	liquid sidestream flowrate	4.3.2
V	fraction of feed that is vapour	4.2.3
V	vapour flowrate	4.3.2
W	vapour side stream flowrate	4.3.2
x	mole fraction liquid	general
y	mole fraction vapour	general
z	total mole fraction in feed	42.

Greek symbols

α	relative volatility	general
Θ	Underwood parameter	general

Subscripts

i	component number	general
F	feed	general
T	overheads	general
B	bottoms	general
MIN	minimum	general
ref	reference component	4.2.4
i, n	stage numbers	general

5 The Flow of Information

A study of the flow of information through a process gives useful insight into the ordering of calculations and indirectly to the whole concept of the design process, which is itself dependent on information flow. We have concentrated on the solution of mass and energy balances for any given process; sometimes called the flowsheeting program.

5.1 Simple mass balances

Many chemical reactions do not go to completion. This means that a percentage of the feed to the process may be recycled from a later stage of the process back to the reactors. The engineer must size the reactors accordingly. A typical sequence is shown in figure 5.1.

5.1.1 SIMPLE ALGEBRAIC SOLUTION

Using the block diagram of figure 5.1 consider a reaction in which benzene and hydrogen are converted to cyclohexane. Assume the reaction goes 80% to completion

$$C_6H_6 + 3H_2 = C_6H_{12}$$

Streams are defined as follows:

$$\begin{aligned} &\text{Feed to process} && = A(1) \\ &\text{Feed to reaction} && = A(2) \\ &\text{Reactor outlet stream} && = A(3) \\ &\text{Product stream} && = A(4) \\ &\text{Recycle stream} && = A(5) \end{aligned}$$

Assume a feed of 100 mol/h of benzene and only consider organic compounds. Once steady state operation is achieved

$$A(1) = A(4) = 100$$

Figure 5.1 Mass balance block diagram.

Considering only benzene to be recycled then

$$\text{Recycle benzene, } A(5) = A(3) = 0.2A(2)$$
$$\text{Also, } A(2) = A(1)+A(5)$$

Substituting for $A(5)$,

$$A(2) = A(1)+0.2A(2)$$
$$\therefore \quad A(2) = 1.25\,A(1)$$
$$= 125 \text{ kg/h}$$

This is an extremely simple example and often plant losses, side reactions, contamination in the recycle, complex kinetics, etc, complicate the simple algebraic solution.

Nagiev (1964) developed this procedure to permit solution for several units, provided the system is linear. Rosen's paper (1962) provides a clear description of the method to use. Care must be taken in interpreting the nomenclature. A three-unit system is shown in figure 5.2, where *each unit is considered as consisting of a mixer node (circle) and a separation module (rectangle) in which vent streams (products) are removed.*

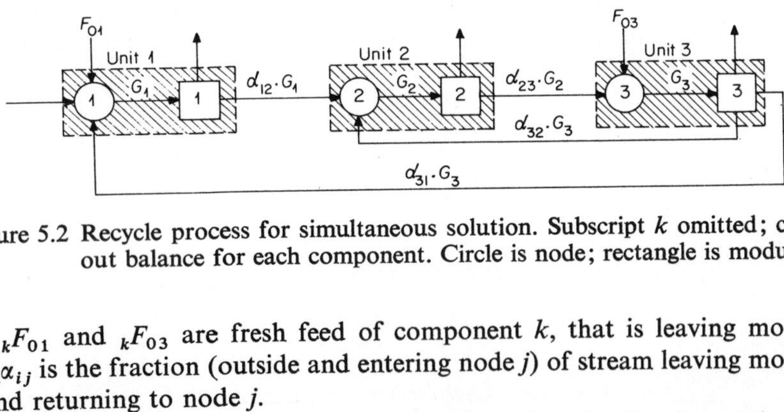

Figure 5.2 Recycle process for simultaneous solution. Subscript k omitted; carry out balance for each component. Circle is node; rectangle is module.

${}_kF_{01}$ and ${}_kF_{03}$ are fresh feed of component k, that is leaving module 0. ${}_k\alpha_{ij}$ is the fraction (outside and entering node j) of stream leaving module i and returning to node j.

${}_kG_i$ is total flow of component k within unit i, i.e. leaving node and entering module. It is the flow rate generated in the computed solution. A material balance at each node (note that the vent/product streams do not enter the calculation) gives for each component

$$\text{Node 1} \quad {}_kF_{01} + {}_k\alpha_{31}\,{}_kG_3 = {}_kG_1$$
$$\text{Node 2} \quad {}_k\alpha_{12}\,{}_kG_1 + {}_k\alpha_{32}\,{}_kG_3 = {}_kG_2$$
$$\text{Node 3} \quad {}_k\alpha_{23}\,{}_kG_2 + {}_kF_{03} = {}_kG_3$$

These may be written as a coefficient matrix of the unknowns ${}_kG_1$, ${}_kG_2$, ${}_kG_3$ and a constant matrix of the input streams ${}_kF_{01}$ and ${}_kF_{03}$.

$$\begin{bmatrix} 1 & 0 & -{}_k\alpha_{31} \\ -{}_k\alpha_{12} & 1 & -{}_k\alpha_{32} \\ 0 & -{}_k\alpha_{23} & 1 \end{bmatrix} \begin{bmatrix} {}_kG_1 \\ {}_kG_2 \\ {}_kG_3 \end{bmatrix} = \begin{bmatrix} F_{01} \\ 0 \\ F_{03} \end{bmatrix}$$

If we know all values of α and $F_{01} \dots F_{03}$ the matrix may be solved by Gaussian elimination as indicated in chapter 3. Note that the system relies heavily on being able to determine the distribution factors α_{ij} and trial and error may be required.

Such systems may be solved using this procedure of carrying out a balance at defined points. Alternatively, a generalized solution may be used in which values of α are substituted directly into a prepared matrix. The latter form is shown in table 5.1. Usually for N process units it is necessary to solve a $N \times N$ system of linear equations for each component.

$$\begin{array}{l} G_1 \dots -a_{11}G_1 \dots -a_{21}G_2 \dots -a_{31}G_3 \dots -a_{n1}G_n = F_{01} \\ G_2 \dots -a_{12}G_1 \dots -a_{22}G_2 \dots -a_{32}G_3 \dots -a_{n2}G_n = F_{02} \\ \vdots \\ G_n \dots -a_{1n}G_1 \dots -a_{2n}G_2 \dots -a_{3n}G_3 \dots -a_{nn}G_n = F_{0n} \end{array}$$

Table 5.1 Linear equations for simultaneous solution. Note: when forming matrix combine $G_1 \dots G_n$ with $a_1G_1 \dots a_{nn}G_n$ to give $(1-a_{nn}G_n)$ term

Let us consider the previous example. If we redraw figure 5.1 just for the benzene balance it is possible to solve the system in one line. Thus from figure 5.3, 20% of unconverted benzene is recycled so $\alpha_{11} = 0.2$. Therefore a balance at node 1 gives

$$G_1 - 0.2\, G_1 = F_{01} = 100$$

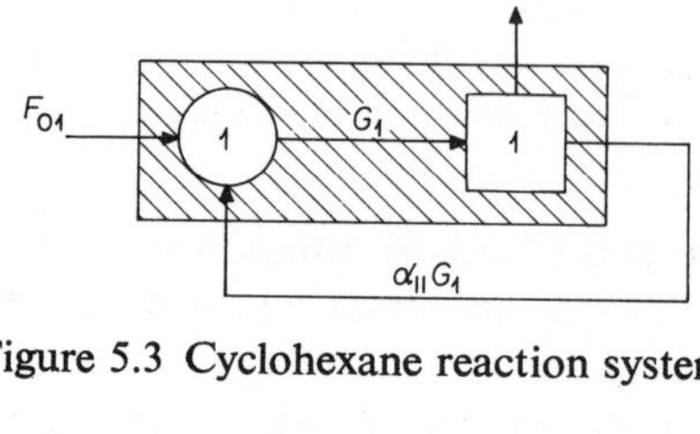

Figure 5.3 Cyclohexane reaction system.

The generalized system assumes links between all possible units and consequently suffers from not utilizing the structure of the network. Our experience with undergraduates suggests writing out the balances to be preferable. For non-linear systems the values of the fractions α_{ij} have to be obtained by a linear approximation. With experience it is unnecessary to draw the node and module as the stream flows are defined within the unit.

Let us consider a more involved example. Mass balances on a toluene hydrocracker: simultaneous solution.

Toluene is converted to benzene by the process

$$H_2 + C_6H_5CH_3 \rightarrow C_6H_6 + CH_4$$

converting hydrogen and producing methane as a byproduct. A molecular ratio of 5 parts hydrogen : 1 part aromatics must be present at the inlet to the reactor. All gas is separated from aromatics in a gravity separator; a quantity of fuel gas required to prevent methane build-up is purged before fresh hydrogen make-up is added. Distillation of aromatics permits recycle of the 2.5% unconverted toluene per pass and 1.5% of the benzene entering the tower is recycled. Ignore traces of toluene in the product.

The mass balances when processing 100 mol/h of toluene are required. The flowsheet is indicated in figure 5.4; see also figure 6.4 for process flow diagram. Using this method of solution it is necessary in order to allow for production of benzene and methane to introduce a dummy reactor with apparent feed of these items. Assume that fresh hydrogen feed consists of 120 mol/h of hydrogen and 7 mol/h of methane.

Using the nomenclature indicated earlier together with subscripts, $T =$ toluene, $B =$ benzene, $M =$ methane, $H =$ hydrogen, so as to permit distinguishing the flows G_1 ... G_5 of the various components, the linear sets may be written as in table 5.2. Further analysis is now required. There are 25

	Linear sets multiply by flow of component in streams					Equate to input
Streams	G_1	G_2	G_3	G_4	G_5	$F_{o,j}$
Toluene balance	1−0	0	0	−1	0	100
	−0.25	1−0	0	0	0	0
	0	−1	1−0	0	0	0
	0	0	−1	1−0	0	0
	0	0	0	0	1	0
Benzene balance	1−0	0	0	−0.015	0	0
	−1	1−0	0	0	0	${}_BF_{0,2}$
	0	−1	1	0		
	0	0	−1	1	0	0
	0	0	0	0	1	0
Hydrogen balance	1−0	0	0	0	−1	0
	$-{}_H\alpha_{1,2}$	1−0	0	0	0	0
	0	−1	1−0	0	0	0
	0	0	0	1−0	0	0
	0	0	$-{}_H\alpha_{3,5}$	0	1−0	120
Methane balance	1−0	0	0	0	−1	0
	−1	1−0	0	0	0	${}_MF_{0,2}$
	0	−1	1−0	0	0	0
	0	0	0	1−0	0	0
	0	0	$-{}_M\alpha_{3,5}$		1−0	7

Table 5.2 Linear set for toluene hydrocracker.

unknown variables and 20 equations. Thus five more equations can be generated as follows:

(1) Benzene produced in reaction

$$_BF_{02} = 0.75\ _TG_1$$

(2) Hydrogen fraction to reactor 2

$$_H\alpha_{12} = (_HG_1 - 0.75\ _TG_1)/_HG_1$$

This modifies the line of the matrix as follows:

$$_HG_2 = {_H\alpha_{12}}{*}{_HG_1}$$
$$= {_HG_1} - 075\ _TG_1$$

(3) Methane produced in reaction, $_MF_{02} = 0.75\ _TG_1$

(4) Ratio of hydrogen to aromatics is 5 : 1 at inlet

$$\therefore\quad _HG_5 = 5(_TG_1 + {_BG_1})$$

(5) Finally the fuel gas purge affects both hydrogen and methane equally so

$$_H\alpha_{35} = {_M\alpha_{35}}$$

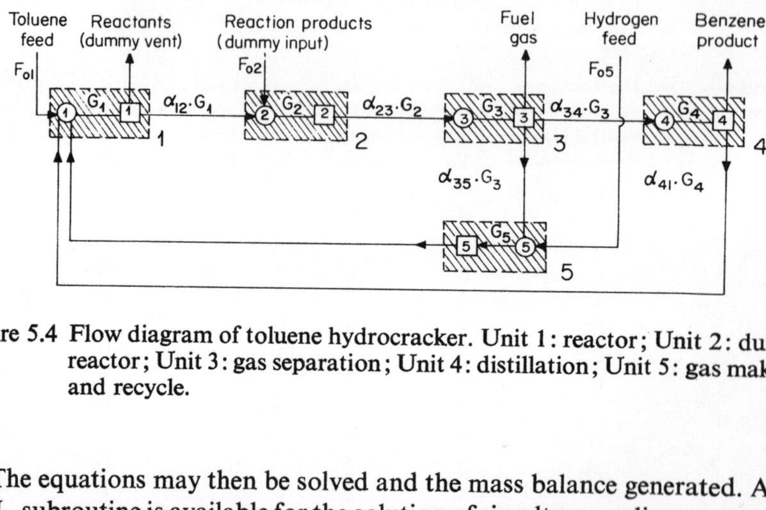

Figure 5.4 Flow diagram of toluene hydrocracker. Unit 1: reactor; Unit 2: dummy reactor; Unit 3: gas separation; Unit 4: distillation; Unit 5: gas make-up and recycle.

The equations may then be solved and the mass balance generated. As an I.C.L. subroutine is available for the solution of simultaneous linear equations the problem was decomposed as in table 5.3. The use of the subroutine is demonstrated in program 5.1 and the solution to the problem is given in table 5.4.

Clearly it is a disadvantage of the procedure, and by no means a simple matter, to tediously develop the equations each time. However, it illustrates a procedure common to many computer-aided designs, that is, generate the equations representing the chemical engineering aspects of the problem and use the appropriate numerical method to produce a solution.

$_TG_1$	$_TG_2$	$_TG_3$	$_TG_4$	$_TG_5$	$_BG_1$	$_BG_2$	$_BG_3$	$_BG_4$	$_BG_5$	$_HG_1$	$_HG_2$	$_BG_3$	$_HG_4$	$_HG_5$	$_kF_{oj}$
1			−1												100
−0.25	1														0
	−1	1													0
		−1	1												0
				1											0
					1			−.015							0
−0.75					−1	1									0
						−1	1								0
							−1	1							0
									1						0
										1				−1	0
0.75										−1	1				0
											−1	1			0
													1		0
5.0					5.0									−1	0

Solve $_H\alpha_{35} = ({_HG_5} - 120)/{_HG_3}$ $\quad {_MF_{02}} = 0.75\,{_TG_1}$

Substitute in following matrix, (5×5), zeros ot shown

$_mG_1$	$_mG_2$	$_mG_3$	$_mG_4$	$_mG_5$	$_kF_{oj}$
1				−1	
−1	1				$_mF_{02}$
	−1	1			
			1		
		$-{_H\alpha_{35}}$		1	7

Table 5.3 Decomposition of matrix for solution by Gaussian elimination. Matrix, (15×15), zeros not shown.

```
      MASTER EQUATION
      DIMENSION A(625),AA(625),B(25),BB(25),X(25),REINT(100)
C   SET UP ARRAY N*N AND VALUES A(I)
      READ(2,4)N
    4 FORMAT(I3)
   15 L=N*N
      DO 10 I=1,L
      READ(2,5)A(I)
    5 FORMAT(F12.6)
   10 CONTINUE
      DO 12 J=1,N
      READ(2,5)B(J)
   12 CONTINUE
      IF(N.EQ.15) GO TO 16
      A(15)=-((X(15)-120)/X(13))
      B(2)=0.75*X(1)
   16 IN=1
      NB=N
      NA=N*N
C   GAUSSIAN ELIMINATION SUBROUTINE OF I.C.L.
      CALL F4ACSL(A,B,N,NA,NB,IN,X,D,ID,IT,AA,BB,REINT)
      WRITE(3,3)IT
    3 FORMAT(1H ,19HNO. OF EXECUTIONS = ,I4)
      DO 11 K=1, N
      WRITE(3,1)K,X(K)
    1 FORMAT(1H ,I3,5X,F12.6)
   11 CONTINUE
      IF(N.EQ.5)GO TO 17
      READ(2,4)M
      N=M
      GO TO 15
   17 STOP
      END
```

Program 5.1 Simultaneous linear equations.

5.1.2 ITERATIONS

Another mass balancing technique is to use iterative calculations until the value of a given stream is within a set tolerance on successive cycles through the system. Thus a guess is made as to the value of the stream and the calculation progresses around the loop until this value can be compared with one system generated.

Consider the simple processing scheme of figure 5.1. Noting that $A(3)$ is representing the cyclohexane content only of this stream, program 5.2 may be used to evaluate the streams in the manner.

As soon as the difference between the initial value of $A(2)$ and the new value of $A(6)$ is within the specific tolerance of 0.000 001 the calculation stops. The alternative to a fixed value is to measure the absolute value of the fractional change ((new value-old value)/(new value)), the system being stated

Number of Executions = 3

1	133.333333	$_TG_1$
2	33.333333	$_TG_2$
3	33.333333	$_TG_3$
4	33.333333	$_TG_4$
5	0.000000	$_TG_5$
6	1.522843	$_BG_1$
7	101.522843	$_BG_2$
8	101.522843	$_BG_3$
9	101.522843	$_BG_4$
10	0.000000	$_BG_5$
11	674.280880	$_HG_1$
12	574.280880	$_HG_2$
13	574.280880	$_HG_3$
14	0.000000	$_HG_4$
15	674.280880	$_HG_5$

Number of Executions = 2

1	2972.402707	$_MG_1$
2	3072.402707	$_MG_2$
3	3072.402707	$_MG_3$
4	0.000000	$_MG_4$
5	2972.402707	$_MG_5$

Table 5.4 Simultaneous linear equations results.

as converged when this is less than a specified tolerance. The number of loops and the values attained before convergence are listed in table 5.5.

The loops are greatly reduced by using a convergence promoter. Such a promoter for the system being evaluated is given by Crowe and coauthors (1971), section 3.1.5. The promoter is introduced as indicated in program 5.3.

```
C  SIMPLE RECYCLE PROBLEM MARK 1
C
C  FLOW RATE OF FLUID ENTERING A1
C
      READ(1,20)A1
   20 FORMAT(F0.0)
      A2=0.
      A3=0.
      N=0
      WRITE(2,6)
    6 FORMAT(1X,10H     N    ,10H     A2     )
   25 N=N+1
      X=A2
      A2=(A1+A3)
      A3=(A2*0.25)
      WRITE(2,30)N,A2
   30 FORMAT(I5,6H      ,F6.3)
C
C  NOW TEST FOR CONVERGENCE
```

```
C
C  DOES REFLUX VARY BY MORE THAN 0.000001
C
      Z=(A2-X)/A2
      IF(Z.LT.0.000001)GOTO 50
      GOTO 25
C  NUMBER OF ITERATIONS IS N
C
   50 WRITE(2,30)N,A2
      STOP
      END
```

Program 5.2 Simple recycle.

```
C  SIMPLE RECYCLE PROBLEM MARK 2
C
C  PROGRAM TO SHOW THAT CONVERGENCE PROMOTER IS
C  MORE ADVANTAGEOUS THAN THE ITERATIVE APPROACH
C
C  VALUES OF VARIABLES AT START
C
      DIMENSION X(20)
      N=1
      READ(1,4)A1
    4 FORMAT(F0.0)
C
C  MASS FLOW RATE OF FLUID ENTERING A1
C
      A2=0.
      WRITE(2,6)
    6 FORMAT(1X,10H         N,10H        A2)
C
C  SIMPLE LOOP TO FIND VALUES OF A2 THE RECYCLE STREAM
C
    7 N=N+1
      A2=A1+A3
      X(N)=A2
      A3=A2*0.25
      WRITE(2,8)N,A2
    8 FORMAT(I5,6H      ,F6.3)
      Z=(A2-X(N-1))/A2
C
C  NOW TEST FOR CONVERGENCE
C
C  DOES RECYCLE VARY BY MORE THAN 0.000001
C
      IF(Z.LT.0.000001)GOTO 11
   10 IF(N-4)7,11,11
C
C  NOW USE CONVERGENCE PROMOTER
```

Program 5.3 Simple recycle with convergence promoter.

```
C
   11 C=(X(N)-X(N-1))/(2*X(N-1)-X(N)-X(N-2))
      Y=X(N)+C*(X(N)-X(N-1))
      WRITE(2,12)Y,C
   12 FORMAT(1H ,F10.3,6X,F10.5)
      STOP
      END
```

Program 5.3 Simple recycle with convergence promoter—continued.

Simple recycle results

LOOP	VARIABLE
N	A2
1	100.000
2	125.000
3	131.250
4	132.813
5	133.203
6	133.301
7	133.325
8	133.331
9	133.333
10	133.333

Simple recycle with convergence promoter

LOOP	VARIABLE
N	A2
2	100.000
3	125.000
4	131.250
133.333	0.33333

Table 5.5 Comparison of convergence.

These results are also listed in table 5.5 and show a considerable reduction in computing time.

The difficulty in using such procedure arises when the system being studied does not converge. Compare the sequence of equations of program 5.3 with a similar series which might have been selected, as follows:

Program 5.3	*Alternative*
→A(2) = A(1)+A(5)	→A(3) = A(5)
A(3) = A(2)*0.2	0.2A(2) = A(3)
A(5) = A(3)	A(5) = A(2)−A(1)
The overall equation is	The overall equation is
A(2) = A(1)+0.2A(2)	0.2A(2) = A(2)−A(1)

Examining the overall equations expressed as

$$y = A(1)+0.2A(2) \qquad\qquad y = 5A(2)-5A(1)$$

and differentiating w.r.t. A(2)

$$f'A(2) = 0.2(<1) \qquad\qquad f'A(2) = 5(>1)$$

If the slope is greater than $|1|$ the solution diverges, see figure 5.5, where values of $A(2) = 75$ and 175 are fed into the two sets of equations. $A(1)$ is taken as 100 in this example.

These criteria is of course only a rough guide and some non-linear systems will converge when the slope is greater than 1. Note that if the equations above are written in matrix form, the L.H.S. variables are on the dominant diagonal, viz

Program 5.3

$$\begin{bmatrix} 1 & 0 & -1 \\ -0.2 & 1 & 0 \\ 0 & -1 & 1 \end{bmatrix} \begin{bmatrix} A(2) \\ A(3) \\ A(5) \end{bmatrix} = \begin{bmatrix} A(1) \\ 0 \\ 0 \end{bmatrix}$$

This matrix is diagonal dominant and will converge.

Alternative

$$\begin{bmatrix} 1 & 0 & -1 \\ -1 & 0.2 & 0 \\ 0 & -1 & 1 \end{bmatrix} \begin{bmatrix} A(3) \\ A(2) \\ A(5) \end{bmatrix} = \begin{bmatrix} 0 \\ 0 \\ -A(1) \end{bmatrix}$$

Note that $0.2 < 1$ for the central row and thus *may* not converge according to the criteria of section 3.2.4.

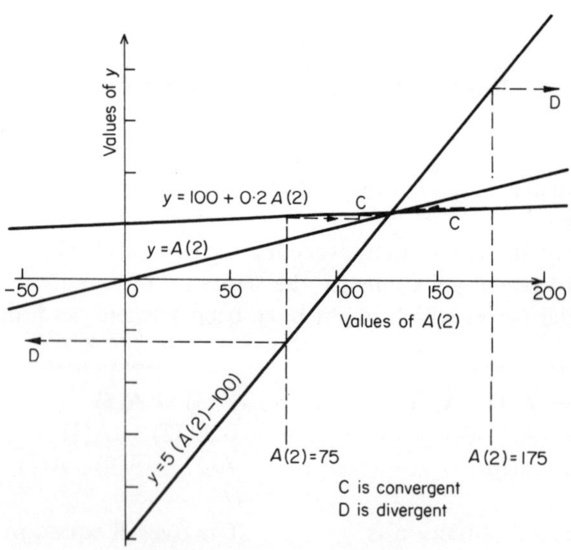

Figure 5.5 Illustration of convergent/divergent behaviour.

Other methods for forcing convergence are available but many of those such as the quasi-Newton methods involve computation of derivatives. They are consequently more difficult to apply and program. Henlay and Rosen (1969) indicate such an example for the Williams–Otto plant.

To illustrate the method tackle the problem of figure 5.2 using an iterative method. This figure is geared to a specific nomenclature and as such the units are difficult to identify in process terms. Rather than use dividers as modules 1 and 3 these have been replaced by distillation columns. Note how module 3 becomes two distillation columns, i.e. to separate a feed stream into three streams conventionally requires two units. Module 2 is taken as a simple divider which does not alter the composition of the streams, i.e. α_{23} is the same for each component. The nodes all become mixer units. Thus the module set consists of:

MIXER, which permits mixing of input streams.

DIVIDER, which permits simple linear split of an input stream into output streams each having the same composition.

SEPARATOR, which distills one input stream according to the specified fraction of each feed stream going to the overhead.

CONTROL, which checks if a specific component stream has converged to a set tolerance.

The units or modules were described as category A programs in chapter 2. The control module is applied to streams and component 2 only in this example. A tolerance of 0.001 is assumed. Let us dimension STREAM (13,2), 13 streams 2 components; MOD (8,2), 8 modules 2 components, X(50) allow 50 loops through the system and set value of X(I) equal to value of STREAM (5,1). Instead of writing the modules out as separate programs it is a simple matter to compute the change in composition for each stream by writing the appropriate arithmetic operation in the correct sequence, program 6.2. The CONTROL module is added and appropriate values for a two-component system read in. Thus values are required for STREAM (1,1; 2,2; 13,1; 13,2) MOD (2,1; 2,2; 4,1; 4,2; 6,1; 6,2). The program can be improved by using better convergence promotion. This method may be written in generalized form only if each module type is called in correct sequence by a master executive and information as to each stream number is conveyed in the array of module parameters.

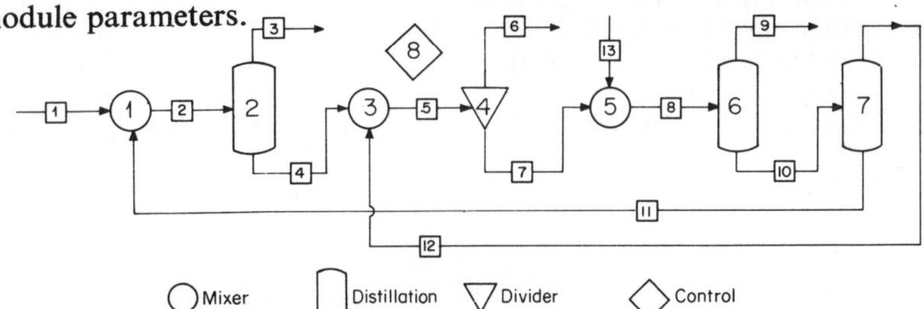

Figure 5.6 Flow diagram of recycle process for iterative solution.

```
      MASTER RECYCLE FLOW
      DIMENSION STREAM(13,2),X(50),F(13,2)
      DO 2 I=1,13
      DO 2 J=1,2
      DO 2 M=1,50
      X(M)=0.
      F(I,J)=0.
      STREAM(I,J)=0.
2     CONTINUE
      READ(1,1000)STREAM(1,1),STREAM(13,1),F(4,1),F(4,2),F(6,1),F(6,2)
      READ(1,1000)STREAM(1,2),STREAM(13,2),F(7,1),F(7,2),F(2,1),F(2,2)
      M=2
3     DO 1 J=1,2
      STREAM(2,J)=STREAM(1,J)+STREAM(11,J)
      STREAM(3,J)=STREAM(2,J)*F(2,J)
      STREAM(4,J)=STREAM(2,J)-STREAM(3,J)
      STREAM(5,J)=STREAM(4,J)+STREAM(12,J)
      STREAM(6,J)=STREAM(5,J)*F(4,J)
      STREAM(7,J)=STREAM(5,J)-STREAM(6,J)
      STREAM(8,J)=STREAM(7,J)+STREAM(13,J)
      STREAM(9,J)=STREAM(8,J)*F(6,J)
      STREAM(10,J)=STREAM(8,J)-STREAM(9,J)
      STREAM(12,J)=STREAM(10,J)*F(7,J)
1     STREAM(11,J)=STREAM(10,J)-STREAM(12,J)
      K=M-1
      WRITE(3,3005)K
      X(M)=STREAM(5,2)
      WRITE(3,3003)X(M)
      TEST=ABS((X(M)-X(M-1))/X(M))
      M=M+1
      WRITE(3,3004)TEST
      IF(TEST-0.001)999,999,3
999   WRITE(3,3002)
3002  FORMAT('0CONVERGED')
      READ(1,1001)
      WRITE(3,1001)
      DO 4 I=1,13
      WRITE(3,3000)STREAM(I,1),STREAM(I,2)
3000  FORMAT(2F10.4)
4     CONTINUE
1000  FORMAT(6F10.4)
3003  FORMAT('STREAM 5,2 =',F10.4)
3004  FORMAT(' TEST =',F20.8)
3005  FORMAT('0LOOP',I5)
1001  FORMAT('SPACE RESERVED FOR SOME SUITABLE TITLE')
      STOP
      END
```

Program 5.4 Recycle with iteration.

5.2 Generalized solutions

Let us consider these methods and see how generalized programs capable of use for many problems can be developed.

5.2.1 ALGEBRAIC APPROACH

An algebraic approach may be used which considers the flowsheet as a set of equations representing the complete process. Algebraic manipulation is then used to reduce the number of equations to be solved iteratively in balancing the recycles. In its simplest form this is a one-off program in which the main effort goes into formulating the problem in such a manner that subsequent optimization of the process is efficient in computation time. The mass balance calculations are kept separate as far as possible from the detailed equipment design, thereby reducing the amount of calculation involved in solving equations which cannot be eliminated.

5.2.2 GENERALIZED ITERATIVE APPROACH

The iterative approach considers the process as a series of interconnected 'black boxes' or building blocks each being a self-contained computational unit. Numerical streams corresponding to the mass and energy flows provide the links and the program permits iteration around the system until steady-state is reached. The black boxes are provided as subroutines to a main calling program, a technique which once developed can be used for many types of processes. The number of loops and hence computing time may be high unless convergence promotion is efficient.

5.2.3 SIMULTANEOUS SOLUTION

Simultaneous solution is recognized as providing extremely rapid solutions. For linear systems the problem is trivial once the matrix is defined. This may be automated using subroutines which describe each unit in terms of an equation, equipment parameters describing the value by which each variable should be multiplied. SYMBOL (Hutchison) is typical of this approach. Once the equation is defined, setting up the matrix is accomplished as follows. Rearranging the columns of figure 5.3 so they appear in stream sequence gives column numbers as

Stream number = I Total components = J						Stream number = I Total components =					
1	2	.	.	.	J	1	2	.	.	.	.
1	2	3	4	5	6	7	8	9	10	.	.

For any value in the equation set, the column number, COL (the lower row identified above), is given by

$$\text{COL} = 1 + \text{I} \times \text{J} - (\text{component identification number})$$

and each row number, ROW, represents the corresponding equation set. The parameter value, PAR, is determined by the module input data. Consider the equation

$$F(1,2) = F(2,2)+5*F(3,2)$$

where in the array $F(i,j)$, i is the stream number and j is the component. For a six-component system

COL(1) = 5	COL(2) = 11	COL(3) = 17
ROW(1) = 1	ROW(2) = 1	ROW(3) = 1
PAR(1) = 1	PAR(2) = −1	PAR(3) = −5

5.2.4 MIXED SYSTEMS APPROACH

For non-linear and linear-mixed systems a general class of iterative methods has been developed in which non-linear functions are linearized. Sargent has given useful reviews. The SPEED-UP approach has been described by Sargent and Westerberg (1964), and Leigh, see Johns (1970). The describing equations for each of the unit operation subroutines are written down in the form $0 = f\{x\}$. The connections between items are specified and the constraints to completely define the system are fixed. The program then performs a logical exercise on the above information, and generates a program to solve the minimum number of simultaneous algebraic equations necessary to define the steady-state of the process. Although the full approach is still being developed, aspects of the philosophy for solving recycle convergence problems are already used in conjunction with other generalized methods. The technique will doubtless grow in use and importance, particularly if computing languages more readily capable of algebraic manipulation become available.

5.3 The basic problem of calculation order

Before attempting the planning of a calculation examine the basic problem of calculation order. First reconsider the case of a single unit. The solution of the set of algebraic equations is only possible if all the input streams and equipment parameters are known. The unit mathematically is composed of sets of N independent equations with M ($> N$) variables. Analysis involves selecting ($M-N$) decision variables to be fixed by the designer (hence input) to leave a ($N \times N$) matrix which may be solved using a method of calculation which can be either simultaneous or involve finding a feasible method by the unit.

If the solution is simultaneous any difficulty lies in formulating and solving the problem. Such a procedure is ideal for carrying out some unit computations. However, for several units at one time unless all the equations are linear, some fearsome partitioning problems may be introduced.

For an individual unit finding a feasible method involves attempting by partitioning to solve each equation in succession such that values are obtained for each variable. The concept is better understood by reference to the overall process. Here the generalized iterative unit by unit method of calculation involves ensuring that the input information to a module is known and the calculated output information can be used as input to the next module. To ensure each module receives satisfactory input information a calculation order must be obtained prior to carrying out the calculation. Most importantly this must be feasible and preferably optimal to reduce calculating time. It may be user supplied to call up subroutines in sequence and the program may check feasibility, or it may be program generated. Much academic interest still centres around whether the user or the computer should provide the calculation order.

In general many problems may be solved using simple graphical tabular representations. Also unlike the computer the user can apply his inbuilt intuition; similar to the chess player's decision made literally 'at a glance' as to which move is the most promising, which the computer has been acertained as requiring 10^{69} seconds to assess. Picasso's dictum 'Je ne cherche pas, je trouve' may be applied.

Strategy of order usually involves finding a simple sequence in which to group units and the objective is to maximize the number of minimum size subproblems. The subproblems are then analysed and a calculation order devised. The usual criterion of this plan, whether applied intuitively or by program, is to minimize the number of tears (values to be assigned to recycle sets), although both Christensen (1970) and Edie (1971) have shown that the minimum does not necessarily mean the optimum solution. As the knowledge of the problem advances the automatic ordering routines may become better able to take account of hidden information flow and adjust the arguments of the convergence routine.

Let us indicate aspects of this strategy.

5.3.1 ASSIGNMENT OF DESIGN VARIABLES

First concentrate on the analysis of calculation order in a single unit. The process design equations can be expressed as

$$f_i(v_j) = 0 \qquad i = 1,2 \ldots N \qquad j = 1,2 \ldots M$$

where f_i is the ith function
v_j is the jth variable
N is the total number of equations
M is the total number of variables.

Inequality constraints also apply, but let us consider only the simpler problem. Rudd and fellow workers have expressed the calculation sequence in the form of a bipartite graph, figure 5.7. The equations form the f nodes, the variables the v nodes and an edge connects the equations to the corresponding variable. The number of edges associated with each mode defines ρ the local degree of the node. The design problem is to assign the decision variables until the logically consistent set of N equations and N variables is obtained. Thus the number of variables to be assigned is equal to F the degrees of freedom of the system, where

$$F = M' - N$$

where M' represents the total number of variables as yet unassigned, i.e. $F = 0$ when $M' = N$.

Consider the following equations as used in figure 5.7:

$$\begin{aligned} f_1\,(v_5, v_2) &= 0 \\ f_2\,(v_1, v_6, v_7, v_8) &= 0 \\ f_3\,(v_1, v_3, v_7, v_8) &= 0 \\ f_4\,(v_1, v_4, v_8) &= 0 \\ f_5\,(v_2, v_3, v_4) &= 0 \\ F = 8-5 &= 3 \end{aligned}$$

Assign three variables as design variables; we will choose v_6, v_7 and v_8. This assigns output sets as marked by the vertical arrows identified by circle

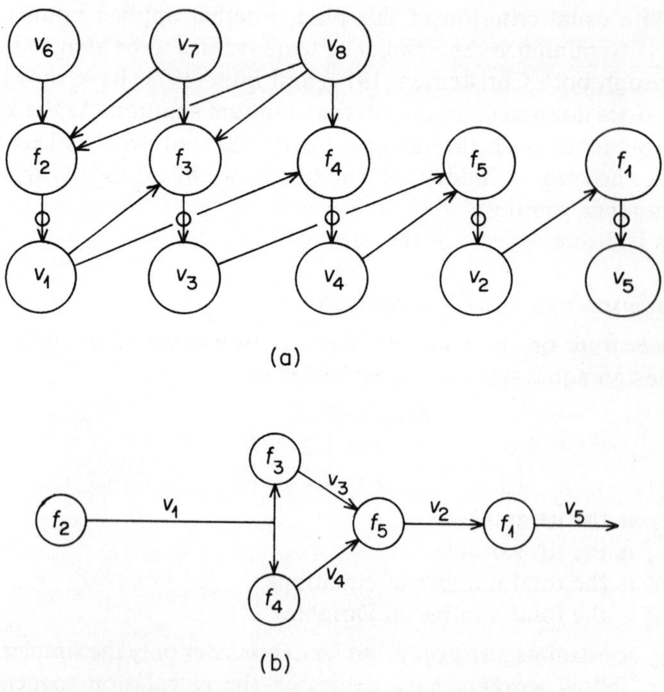

Figure 5.7 Graph of equations, (a) Assignment of output sets. (b) Precedence order.

and an order of solution is suggested by the network. The precedence order is shown clearly by rearranging the graph and in this case has an acyclic structure, i.e. it has no path which returns to its starting node. If the system does possess a loop, the simultaneous solution of those equations within the loop must be undertaken.

One criterion for the selection of design variables is that a best set of design variables result in a structure which minimizes the number of equations that must be solved simultaneously. Clearly an acyclic structure is the best solution. It represent the most rapid computational scheme. The graph of such a system must contain one v node v_j with $\rho(v_j) = 1$ and one f node f_i with $\rho(f_i) = 1$, because every directed path in an acyclic graph must terminate by definition at a v node. Also by definition a v node may only have one incoming edge. So a v node must have exactly one edge if the path terminates at that node.

Rudd and coauthors applied the algorithm, table 5.6, to such graphs. If $\rho(v_A) \geqslant 2$ for a remaining v_A after applying the algorithm, such a system

(1) Set out the structural array of f equations (rows) and v variables (columns).
(2) Locate a column containing only one variable.
Delete that column and corresponding row.
(3) Repeat until all equations are deleted.
If the above steps fail to terminate continue as follows.
(4) Define $k = \min \rho(x_i - 1)$ where $\rho(x_i)$ is the number of equations in which x_i appears.
(5) Identify sets of k equations such that when the set is deleted step 2 may commence.
(6) Delete one such equation set and commence at step 2.

Table 5.6 Variable selection algorithm (Rudd and Watson, 1968)

is not acyclic and the flow can only be made acyclic by 'tearing' some of the state variables, i.e. an output is assigned to k f nodes to permit their deletion from the graph. Then an acyclic system may be possible. The objective is to solve the system with minimum tears. As it is usually more convenient to work in tabular form, this is based on a matrix version of these graphs.

Ordering with minimum tears

Let us consider two small systems in which the equations and variables are defined and order the calculation sequence according to the principle that if 'tears' are required they will be the minimum number for the system. The system to be analysed was considered earlier in this section. The application of the algorithm is given in table 5.7 and the same solution as given in figure 5.7 is deliberately contrived. The discerning reader will note that many combinations of the variables could have been selected as design variables in this case.

f/V	V_1	V_2	V_3	V_4	V_5	V_6	V_7	V_8
1		*			*			
2	*					*	*	*
3	*		*				*	*
4	*			*				*
5		*	*	*				

Choice—delete V_5 or V_6, say select V_5 and delete equation 1

	V_1	V_2	V_3	V_4	V_6	V_7	V_8
2	*				*	*	*
3	*		*			*	*
4	*			*			*
5		*	*	*			

Choice—delete V_6 or V_2; say select V_2 and delete equation 5

	V_1	V_3	V_4	V_6	V_7	V_8
2	*			*	*	*
3	*	*			*	*
4	*		*			*

Choice—delete V_3, V_4 or V_6; say select V_4 and delete equation 4

	V_1	V_3	V_6	V_7	V_8
2	*		*	*	*
3	*	*		*	*

Choice—delete V_3 or V_6; say select V_3 and delete equation 3

	V_1	V_6	V_7	V_8
2	*	*	*	*

Choice—delete V_1, V_6, V_7 or V_8; say select V_1.

Then V_6, V_7 and V_8 are the design variables and the precedence order or calculation is the reverse of the sequence of deletion. This solution is the same as noted earlier in figure 5.5, i.e.

(1) Obtain values of V_6, V_7, V_8 for the design variables.
(2) Solve equation 2 for V_1.
(3) Solve equation 3 for V_3.
(4) Solve equation 4 for V_4.
(5) Solve equation 5 for V_2.
(6) Solve equation 1 for V_5.

Table 5.7 Precedence order of equation set with no recycle.

The system defined in table 5.8 does not reduce so fortuitously and in fact persistent recycle occurs. One of the variables must be selected as a 'tear' variable. Details of the analysis and one of the possible solutions to the problem is given in this table.

It is self-evident that carrying out analysis of small systems in this fashion is of considerable help in ordering programs. Variables can be selected which lead to the greatest ease in solving the problem. Also it is possible to check whether the variables specified permit solution of the problem. A recommended exercise is to consider the overall mass balance equations for a distillation system and so determine the number of variables to be specified.

f/V	V_1	V_2	V_3	V_4	V_5	V_6	V_7	V_8
1		*			*			
2	*				*	*	*	*
3	*		*			*	*	*
4	*			*				*
5		*	*	*				

No initial selection is possible.
For variables V_2 is possible.
For variables V_2 and V_5, $k = 2-1 = 1$
Delete equation 1

	V_1	V_2	V_3	V_4	V_5	V_6	V_7	V_8
2	*				*	*	*	*
3	*		*			*	*	*
4	*			*				*
5		*	*	*				

Delete V_2 or V_5; say V_5 and equation 2

	V_1	V_2	V_3	V_4	V_6	V_7	V_8
3	*		*		*	*	*
4	*			*			*
5		*	*	*			

Delete V_2, V_6 or V_7; say V_6 and equation 3

	V_1	V_2	V_3	V_4	V_7	V_8
4	*			*	*	*
5		*	*	*		*

Delete V_1, V_2, V_3 or V_7; say V_1 and equation 4

	V_2	V_3	V_4	V_7	V_8
5	*	*	*		*

Delete V_2, V_3, V_4 or V_8; say V_3.

Of the four remaining variables as $k = 1$, one is a recycle variable V_2 and V_4, V_7 and V_8 are design variables.

(1) Assume V_2.
(2) Obtain values of V_4, V_7, V_8, the design variables.
(3) Solve equation 5 for V_3.
(4) Solve equation 4 for V_1.
(5) Solve equation 3 for V_6.
(6) Solve equation 2 for V_5.
(7) Solve equation 1 for computed value of V_2.
(8) Repeat until assumed and computed values of V_2 are the same.

Table 5.8 Precedence order of equation set with recycle.

5.3.2 INFORMATION FLOW DIAGRAMS

The algorithms used to aid analysis of the design equations are adapted to a modified form of the process flow diagram, termed the information flow diagram. This is a directed graph which consists of a collection of nodes linked by edges. If more than one edge exists between nodes i and j a weighting w_{ij} may be given. The process units represent nodes and the edges the process parameters. Now the desired computational scheme is acyclic. Thus the units of the plant are arranged into a sequence of units or groups of units such that there is only a forward feed of information from group to group. However, individual units may contain many self-loops and further analysis can be required to break down these recycle streams. Indeed it may be desirable to represent the equipment modules as subnetworks to take full advantage of the ordering facilities. This is, however, for very advanced ordering systems.

For design purposes it should be noted that *the flow of information is not necessarily in the same direction as flow of material*, i.e. the specification of a product stream means a flow of information to a process unit as opposed to the physical flow of the stream out of the unit. This is important as it represents input information to the computer. The $(M-N)$ decision variables are assigned by the environment, e.g. mains steam pressure, cooling water temperature; others are invariant by the statement of the problem, e.g. selling specification for product; and those remaining are consumed in maximizing the profitability of the process. Free variables are consumed upon linking the individual systems together. The resulting sturcture is examined. Figure 5.8(a) shows one method of decomposing the cycle. In this case calculations are carried out until the values match, i.e. they have converged.

Alternatively reversal of flow may be feasible. In figure 5.8(c) flow has been reversed which means a change of direction of dependent computed variables round the system until a suitable free variable at which to reverse flow is obtained. Note how there is no change in the number of arrows providing information into each unit. These represent the $(M-N)$ decision

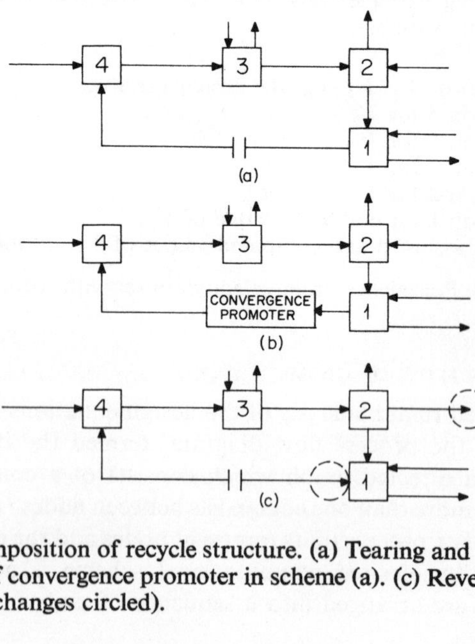

Figure 5.8 Decomposition of recycle structure. (a) Tearing and iteration. (b) Insertion of convergence promoter in scheme (a). (c) Reversal of information flow (changes circled).

variables which are fixed in number. Note that in figure 5.8(c) the technique employed in optimizing the system is to set the parameter at various values and check the effect on the objective function. The system is acyclic. In the system of figure 5.8(b) the objective is agreement between assumed and computed values. The system is a recycle set. In this case improved iterative procedures are desirable as successive substitution can take excessive computing time. Both systems may be regarded as forms of 'tearing' if this is defined as guessing the value of an edge so as to cut an existing loop. The identification of recycle sets is the most used strategy. Then the use of a convergence promoter to test and force convergence is possible, as indicated in figure 5.8(b).

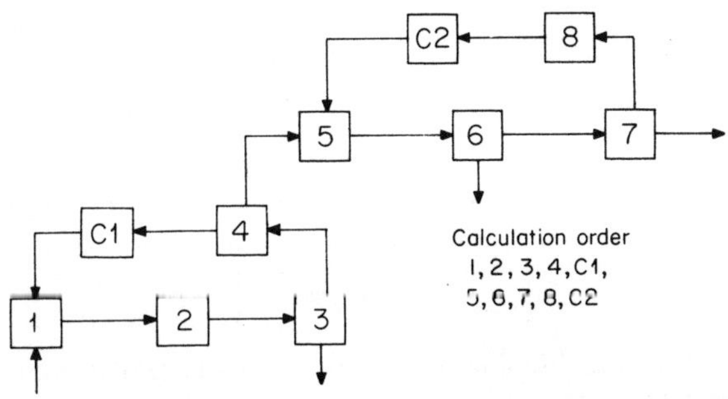

Figure 5.9 Simple partitioning.

5.4 Planning the process calculation

The planning of the process calculation is defined by Crowe and coauthors (1971) as

(a) Separating the units into those which form part of recycle loops and those which may be directly calculable.

(b) Planning the calculation of the separate recycle sets.

Such planning may frequently be carried out by simple inspection, followed by indicating a calculation order to the computer. The following considerations give some experience in formulating this approach.

5.4.1 PARTITIONING

Let us for the moment consider each unit computation as a building block. The convergence promoters are additional blocks which are added at the user's discretion. As indicated earlier to solve a set of $N \times N$ equations involves partitioning the equations into the smallest set of equations which include all the variables which must be solved simultaneously. These building blocks may be divided into similar groups. Such partitioning is frequently obvious and careful redrawing of the system suffices. Thus as in figure 5.9 the system may be partitioned into two clear sets. A convergence promoter,

C, has been added into both loops or alternatively simultaneous solution of the two systems may be used. Figure 5.10 is a more difficult network to decipher and is here partitioned into three systems.

5.4.2 NESTING

Having obtained independent groups a strategy of calculation must be devised. Consider the sequence on the information flow diagram figure 5.11. We can plan several strategies for evaluating this system. Streams are labelled alphabetically and modules numerically.

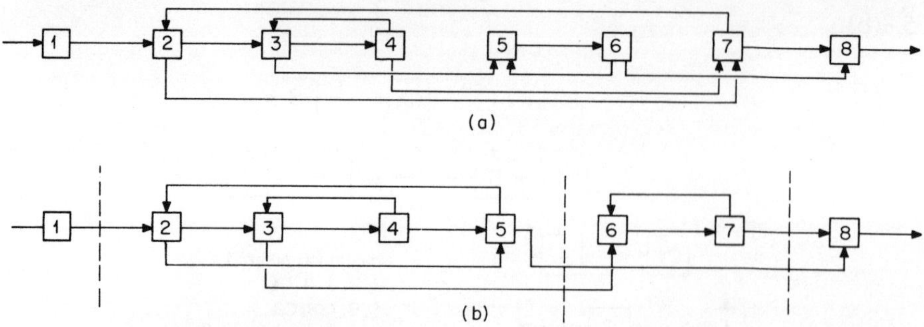

Figure 5.10 Result of partitioning. (a) Network before partitioning. (b) Network after partitioning .

(a) For the external loop 1, 2, 5, 1 a guess could be made of stream *f*, which permits streams *a*, *b*, *g* to be computed. For loop 3, 4, 3 an estimate of stream *d* could be made and following computation of stream *c* an alternative estimate made. The discrepancies between the initial estimate of flow *d* and the revised estimates may be used to define a set of simultaneous non-linear equations for the flows *c* and *d*. The iterative solution of these equations gives

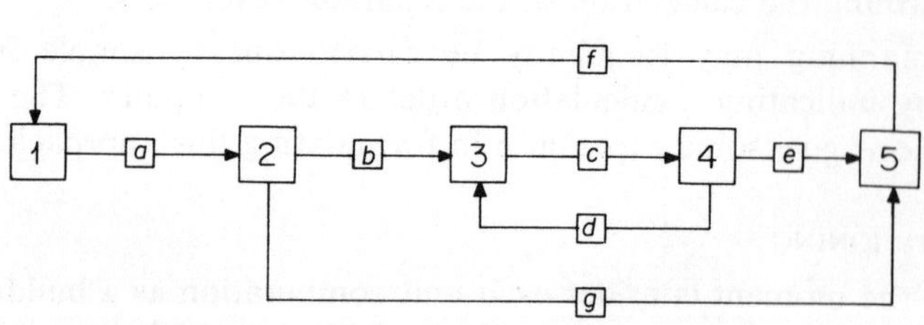

Figure 5.11 Nesting recycle streams.

a balanced set of flows in the loop 3, 4, 3 which can be used to compute stream *e* and hence a revised estimate of flows in stream *f*. The set of equations for stream *f* are defined as for stream *d*.

This his approach requires the solution of the set of equations to solve for

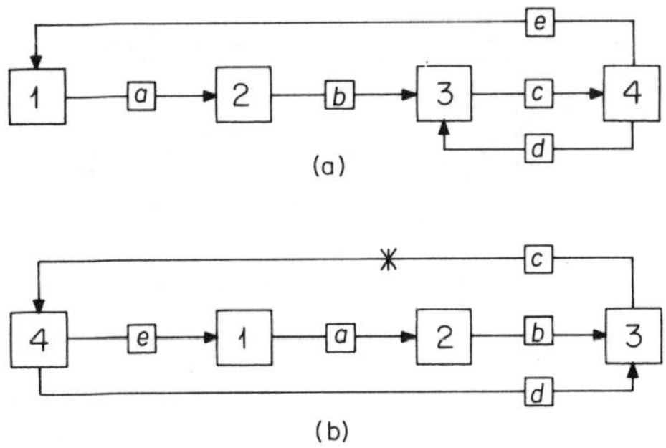

Figure 5.12 Selection of tear streams. (a) Before selection. (b) After selection, redrawn.

d at every step of the iteration required to obtain *f*. The technique is known as 'nesting' the solution of the loop 3, 4, 3 within the loop 1, 2, 5, 1.

(b) Secondly, we could adopt a 'simultaneous' solution by making initial estimates of *d* and *f* permitting *a*, *b*, *g*, *c*, *e*, to be computed and thence revised estimates of *d* and *f*.

Other streams could have been selected as 'tear' streams. Johns (1970) has discussed the alternative strategies and demonstrates that at the present state of knowledge the 'simultaneous' solution should normally be employed for economic reasons. However, special routines are required to apply such technique, and the less fortunate user may therefore be forced to use the fully iterative approach. This approach may be used for this problem by inserting convergence promoters into each loop. Rather than redrawing each diagram let us identify the sequence of calculation by a calculation order list. Alternatives are

1 , 2 , 3 , 4 , *C* on *d* , 5 , *C* on *f*

and

1 , 2 , 3 , 4 , 5 , *C* on *f*

5.4.3 MINIMUM NUMBER OF RECYCLE SETS

Identifying the minimum number of recycle sets is possible on the information flow diagram, but beware of instant guesses.

Consider figure 5.12; at first glance it might be decided that two recycle streams *d* and *e* must be studied for convergence. However, if stream *c* is treated as the recycle stream the problem is simplified as the rearrangement shows. Note that strictly speaking the aim is to minimize the total number of variables in the cut streams so weighting values on each stream would normally be incorporated as in the following example. Figure 5.13 identifies the recycle streams to tear. If simple inspection is not satisfactory, there are many algorithms for ordering the calculation, and specialist knowledge of this field

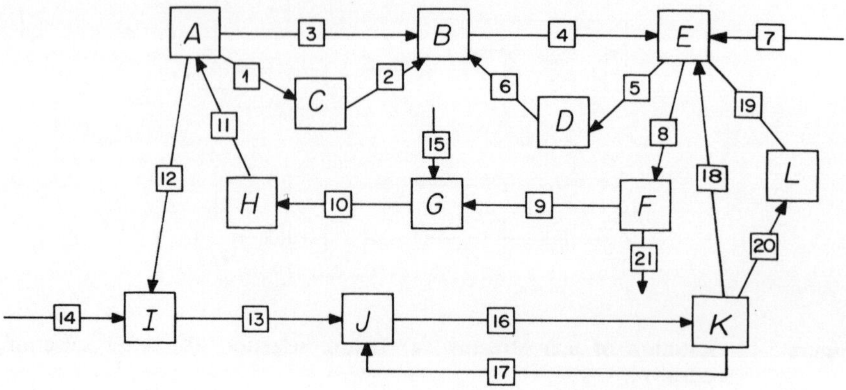

Stream no.	Variables	Stream no.	Variables	Stream no.	Variables	Stream no.	Variables
1	3	6	2	11	3	16	5
2	3	7	2	12	3	17	1
3	3	8	3	13	5	18	5
4	3	9	2	14	3	19	3
5	3	10	3	15	1	20	2
						21	2

Figure 5.13 Selection of streams for convergence promotion.

may readily be obtained. Valuable insight is given by Sargent and Westerberg (1964), Rudd and Watson (1968), Christenson (1970), Himmelbau (1968) and Johns (1970). The systematic search program developed by Shannon is given by Johnson (1970) and is interesting to use as a check on simple inspection methods to find the minimum number of recycle streams. The program of Christenson is available in his Ph.D. thesis.

It should be noted that these workers recognize that partitioning and ordering may produce computation schemes which are inefficient and unstable, but at the same time their approach provides useful insight into the structure of the process. More recently the problem for the complete system is being considered in its mathematical origin as one of solving a set of N equations with M variables and selecting the $(M-N)$ decision variables in such a manner as to assign the output set based on covergence characteristics. The remaining $(N \times N)$ matrix is solved simultaneously. The size of the problem can be overwhelming for very large systems.

Notation	Definition	Section
$A(I)$	stream identifier	general
a–g	stream identifiers	5.4
F	feed flowrate	5.1
F	number of degrees of freedom	5.3
$F(i, j)$	flowrate array	5.2

f	function	5.3.2
G	component flowrate	5.1
i, I	stream number	5.2
j, J	component number	5.2
k	set of equations	5.3.1
M	total number of variables	general
N	total number of equations	general
N	number of process units	5.1
v, V	variables	5.3

Greek symbols

α	distribution factor	5.1
$\rho(x)$	number of equations involving x	5.3.1

Subscripts

i	stream number	5.2
i	module number	5.1
j	component	5.2
j	node number	5.1
k	component	5.1
B	benzene	5.1
H	hydrogen	5.1
M	methane	5.1
T	toluene	5.1

Superscripts

$'$	first derivative	5.1.2
$'$	total number of as yet unassigned variables	5.3.1

6 Process Simulation

6.1 General

A process simulation can be used in the development and design of a new process or to simulate an existing plant. A process simulator or generalized flow-sheeting program makes this possible on the computer. The operations which the simulator can most readily evaluate are indicated on the greatly simplified project network in figure 6.1.

A procedure to follow in order to obtain the most economic plant design is given by Beckett (1969) as:

(a) Set the product tonnage and purity, and the practical process stages.

(b) Prepare a flow diagram and define the constraints (pressure, temperature).

(c) Calculate the flow rate and physical conditions of each stream and the utility requirements.

(d) Design the process units.

(e) Estimate the capital cost of the main items of plant and estimate the total installed cost.

(f) Determine operating costs.

(g) Calculate the net present worth of the plant.

The procedure is cyclic being followed several times until an optimum design is reached.

As the simulation is required at various stages of the design or operation of the plant a generalized solution must have flexibility. How may this be achieved? We have seen that a chemical engineering system consists of a network of equipment modules in which various physical or chemical transformations take place under the influence of variations in heat content, pressure and chemical activity. Note that by equipment is meant not only major items such as columns and reactors but also valves. Also a piece of equipment can be modelled in various ways (even by several modules) according to the requirements of the engineer. For example, as indicated in chapter 2, various distillation subroutines can be used as follows:

Category A The subroutine may carry out a simple linear split of one input stream into the overhead and bottoms streams. The approach is sufficient

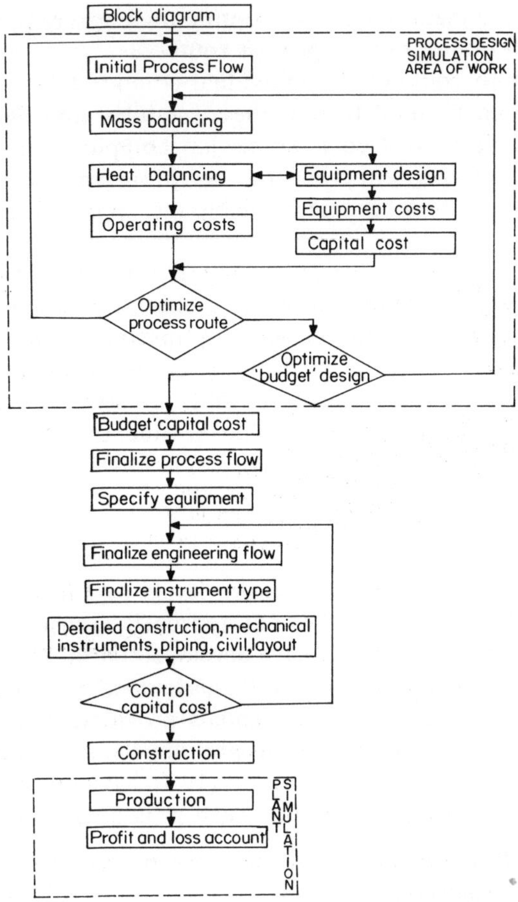

Figure 6.1 Simplified project structure emphasizing simulation.

for simulating existing plants which are to be controlled to rigid specification or for obtaining mass balances at an early stage of the design.

Category B A short cut distillation procedure such as Fenske–Underwood–Gilliland can be used to provide a rating/design option. This procedure requires less data input and is quick to converge. Thus its use may save time in the early stages of a design if many iterations through the system are anticipated, i.e. optimization of process route.

Category C A major distillation program as indicated in chapter 4 may be incorporated. Our experience has suggested that a preferred procedure is to use these as self-standing programs to update and check Fenske values. Otherwise unless the self-standing program converges rapidly and the physical properties package is extremely adequate, a lot of preliminary formulation plus computing time and storage is involved in using the simulator. Our suggestion is to use them when a near optimum design requires finalizing but not for general optimization of process route.

It is feasible to consider the individual column trays as modules. The executive can then be used to link these modules and design the system. The recycle problem involved is somewhat complex and at present such programs remain largely experimental. As programs solving the problem by simultaneous solution become available the approach may have greater value.

It should be noted that in the design stage it is necessary to use some highly sophisticated design programs, either as a check on simpler routines used because of speed of computation when optimising, or as an essential and separate stage because of the inadequacy of the 'flowsheeting' simulator. Such applications may be proprietary and only available for the use and at the offices of specialist designers. Thus a function of the simulator is to provide some of the input data to these programs.

6.2 Process simulators

Several executive systems exist and are reviewed by Flowers and Whitehead (1973). Generally these are available at appropriate cost although some are retained by the companies which have developed them. *It is essential to have access and use a system in order to fully understand the advantages and disadvantages involved.* Although the manuals provided with the systems are easy to comprehend and give good insight into their operation, in this field practice is essential. Crowe and coauthors (1970) have given some insight into the problem. The key aspects are that such programs should be

Modular: permitting users to insert their own procedures.
Openended: for expansion.
Flexible: according to the stage reached by the design.
Readily understood and usable: by other than experts in this area.

The operation of these simulators is as follows:

INITIALIZATION involves input of a list of the subroutines for equipment modules, indicating the links between these streams, checking feasibility of sequence of calculation, input of parameters and initial stream values.

The EXECUTIVE program calls the data reading subroutines to load the input data. It then manipulates this data in an execution sequence which frequently involves an iterative calculation in which the unit computations modify input streams to produce output streams. The executive program usually activates the supply of output information, either directly from the executive or through an interfaced *report writing subroutine.* Such information is optional but for design purposes should include all normal process flow diagram information such as numbers and names of streams and components, physical state of streams including temperature, pressure, fraction vapour/liquid/solid, flow rates both weight and molar. Further information such as sizes and costs of items, raw material and utility consumptions and costs, profitability in appropriate form is desirable. Process specification of units should preferably be in a form suitable for submission to fabricators for tender or designed to provide ease of input to a category C program.

Executive subroutines are used to assist in manipulating information. We have shown earlier how major unit computations may be built up out of subroutines. Basically these comprise the equations representing the fundamentals of the unit operation and a second group representing means of algebraic manipulation in order to solve the former equations. Clearly if the two can be kept somewhat separate this eases the formulation of new subroutines representing equipment modules.

The simulator operates in a similar manner to a unit computation. The following loose classification indicates the routines considered desirable.

Printout Built into the program should be IF statements which permit large amounts of printout from key steps of the program when debugging: warnings and error messages, etc. The executive subroutine can specify which data is required.

Ordering Partitioning, simplifying and calculation as indicated in chapter 5 may be automated or programmed separately as an order for calling modules.

Numerical methods Such routines, in particular for convergence promotion, may vary from simple application to a single stream in an equipment module to solving equations simultaneously for several streams. They might incorporate optimization subroutines as indicated later in this chapter.

Changing data Control duties can reset parameters such as flow of a single component, a whole stream or loop numbers. They can transport information from one module to another. They enable limitation of the simulator due to its necessary generalization to be overcome. A further use is in permitting more than one set of input data to be processed per run.

Files of data may be stored in the computer. Physical properties data may be provided separately to each subroutine but is more conveniently provided by a package containing a range of generalized correlations of thermodynamic and transport properties and the means of evaluating these properties for each state of stream etc, as described in chapter 2. Further data for this package may be accessed from a major data source such as the NEL–APPES system described by Martin (1972). Also it is useful to store cost data from previous projects.

The use of *module subroutines* has been indicated earlier and their use depends on the objective function of the specific task. It should be appreciated that many of these subroutines are used for modules at several points of the system and this greatly reduces the number of specific module subroutines which have to be supplied by the user.

It is preferable that if the routines contain a rating option they include correlations for, or access to, data for estimation of capital cost. This permits more ready optimization of process route. Provision is desirable to permit evaluation of installed from delivered costs.

6.2.1 CLASSIFICATION

With such a range of routines from which to select it is desirable to attempt some classification to assist the user in selection. Again the categories A, B and C will be used as some correspondance with program classification is possible. Table 6.1 indicates the range of routines which might be incorporated.

Routine	Category A	Category B	Category C
Executive	Executive	Executive	Executive plus data check and zero initialization routine.
Print out	Options on input and internal. Final output streams.	As A plus rough rating of equipment and capital and operating costs.	As B plus specification sheets for equipment and net present value.
Calculation order	User supplied.	User supplied.	Ordered automatically.
Unit computations	Category A.	Categories A and B.	Categories A, B and C.
Convergence	Wegstein or simultaneous.	As A for mass balance, iteration for heat balance, internal module routines.	Decomposition into minimum recycle sets which are solved simultaneously. Package of convergence routines.
Thermodynamic	User supplied in subroutines.	Small package evaluating approximate properties. Input basic data.	Comprehensive package with access to large data store.
Capital cost	None.	Correlations, accuracy only $\pm 20\%$.	Data package and algorithms.
Optimization	External, generate more data by alteration routine.	By user through data alteration. Aim to optimize process route.	Optimization algorithms. Optimize route and equipment.

Table 6.1 Classification of steady-state process simulators.

Category A Mass balancing

This range of simulators is of some value in the simulation of existing plant due to their use as building blocks. But clearly one-off programs also may easily be written. Their more important role is in training as they provide instruction in method. It is useful to study both iterative and simultaneous solution, e.g. GEMCS and SYMBOL. It is important that such simulators occupy relatively small core, converge rapidly and contain data alteration routines for carrying out several runs per batch job. The use of automatic calculation ordering is superfluous.

Category B Optimization of process route

There is a need for simulators capable of carrying out optimization of process route. Such programs should be capable of giving a rough rating of capacity, capital and operating costs. However as the rating is approximate the thermodynamic package should be kept relatively simple and complex exchanger trains lumped together. As will be indicated later in this chapter we favour analysis of information flow and ordering of calculation sequence by the user. It is most important that such programs be readily understood and usable by non-experts. If errors occur ample diagnostic checks should be given so that the user may readily rectify errors, particularly if occurring in computation of physical properties.

Category C Comprehensive simulation

The package to optimize both route and equipment is best developed by firms for their specific use. The thermodynamic package is then geared to their requirements. Optimization algorithms can be developed for repeating problems and due allowance made for subsequent use of self-standing category C programs. The size and complexity of such simulators tends to make them difficult to transfer between different machines and the user has to become very familiar with the capabilities of the program before errors and consequent expensive computing costs are avoided.

6.2.2 COMBINED SYSTEMS

A useful approach is combining the separate categories of simulator and to follow the logic of the process engineer as he goes about a similar task using time-honoured procedures. He gradually builds up the complexity of the system, first using simple means and refining these as the possibility of sanction of capital expenditure increases. A similar approach is possible using simulators and the approach of Klumpar (1969) seems the closest to a desirable procedure. We will assume the project has got past the preliminary evaluation stage and the aim is to produce a budget estimation. It is assumed that there is a high probability of the plant going ahead and that it is sufficiently novel for conventional estimating techniques based on similar plant to be used.

(a) For mass balancing use a category A simulator to achieve material balances. The output from this simulator can if necessary be in the form of punched card input to the second stage which is

(b) Energy balances, rating and costing of process equipment using a category B simulator. Energy balances and equipment rating cannot be divorced. Heat of reaction, reboiler duties all affect heat exchanger requirements. At the same time it is preferable to cost equipment in the routine carrying out the rating. It is probably the one item here which causes any difficulty is determining the material of construction before thermal conditions are defined.

(c) The investment may then be estimated based on delivered equipment cost. Operating costs should be included in this section as report writing is thereby improved. One has the option of using all the programs together if one is confident of achieving a successful end product—capital and operating costs for a specific capacity; or it is possible to develop an answer for one specific case and until such time as this is proven use the program in sections. Such a system should result in lower computation costs and in addition, where computation time per batch run or core size is limited, permit the use of a large simulator.

As will emerge later our view is that for all but those people fortunate enough to have a highly sophisticated and reliable simulator the use of program order is not necessary. Also we like to partition completely the mass and heat balancing; so much so as to use a different calculation order in these stages. It will be appreciated how the above system lends itself to this approach.

6.3 PRIMER

As an example of a category A simulator let us briefly discuss the simulator PRIMER (*Process Iterative Mass-Flow Evaluator*) developed by Burgess, Robson and Wells at Sheffield University. This program is a modification of the GEMCS system originally devised by Johnson (1970) at McMaster University. The PRIMER system allows for the mass and user supplied heat balancing of steady-state chemical plant and will also permit an economic analysis of an existing plant using the same arrays along with costing subroutines.

The two major arrays in the system are:

FLOW which stores all plant streams permanently,

UNIT which stores all module parameters.

The dimensions of these arrays are fixed and data is stored as far as possible in a standard order. PRIMER defines the FLOW array in reverse to convention with the stream number second.

J	FLOW (J, I)—I is the stream number.
1	Stream number.
2	Mass flow.
3	Stream name code number or enthalpy.
4	Cost/unit.
5	Temperature.
6	Pressure.
7	Flow of component 1.
8	Flow of component 2.
.	
.	
.	
JJ	Flow of component NCOMP.

Table 6.2 Flow array format.

The FLOW vector format is given in table 6.2. Consequently each stream vector consists of six numbers and all the component flows. It is essential that all streams vectors have the same length, JJ, which in the present version is equal to 20, so permitting a system of fourteen components. The fourth element is given a value when carrying out economic simulation. Elements 5 and 6 are optional when carrying out a mass balance simulation. In every stream the flow of each component (even if zero) must be included. This applies particularly to the input streams where data is supplied by the user.

N	UNIT (N, I)—I is the module number.
1	Module number.
2	Module type.
3	Number of input streams to module.
4	Number of output streams from module.
5 . . . 20	Input and output stream numbers.
21 . . . 30	Equipment parameters peculiar to specific module.

Table 6.3 Unit array format.

The UNIT array format is given in table 6.3. As with the stream vector format the length of the UNIT vector is standardized being equal to 30. This allows for a total of sixteen input and output streams from any one module and up to ten equipment parameters for a specific module. The other important program variables are

LIST(50)	—calculation order of modules allowing a maximum of 50 unit subroutines.
NCYCLE	set = 0 for no recycle, set = 1 for recycle.
LOOP	set = 999 when no recycle in system—this is equivalent to the allowed maximum number of loops before termination of execution.
NIN	number of input streams.
NOUT	number of output streams.
NCALC	number of modules in LIST.
NFLOW	total number of streams.
NUNIT	total number of modules.
IFLOW	number of initial streams read.
MLOOP	reset LOOP number after data modification.
NU	order of calculation subscript for calling of modules subroutines.
KPRINT(3)	control of output printing.
NRUNS	number of runs equal to first case plus number of modified cases.
MRUN	modified run counter.

The data input is achieved using the NAMELIST format available in extended Fortran. This method has been found to be more desirable, in terms of ease of writing data cards, than the approach of using the same field width format for all inputs via standard READ statements. An example of the NAMELIST method as used in PRIMER is

```
   NAMELIST/CLIST/NCYCLE,NCALC,NCOMP,LIST,NFLOW,
  1 IFLOW,NUNIT,KPRINT.
   :
   READ (X,CLIST)
```

The corresponding data cards would be

```
—&CLIST
—NCYCLE=X
—NCALC=X
—NCOMP=X
—LIST=X,X,X, ----   up to LIST(N) value
—NFLOW=X
—IFLOW=X
—NUNIT=X
—KPRINT=X,X,X
—&END
```

In order to permit the storage of data, where X represents numbers, using NAMELIST, in the double dimension arrays FLOW (J,I) and UNIT(N,I), it is necessary to set up dummy arrays as shown:

COMMON/UNITS/UNIT(30,50) in EXECUTIVE
COMMON/FLOWS/FLOW(20,100)

and,

COMMON/UNITS/UNIT1(30),UNIT2(30), ... UNIT50(30)
COMMON/FLOWS/FLOW1(20),FLOW2(20), ... FLOW100(20)

in DLOAD data reading routine.

The NAMELIST cards have the following format,
NAMELIST/UNITLIST/UNIT1,UNIT2, ... UNIT50
NAMELIST/FLOWLIST/FLOW1,FLOW2, ... FLOW100

the data cards are,

```
—&UNITLIST
—UNIT1=1,X, ...   up to UNIT1(30) value.
—UNIT2=2,X, ...   up to UNIT2(30) value.
etc


—&END
—&FLOWLIST
—FLOW1=1,X, ...   up to FLOW1(20) value.
—FLOW2=2,X, ...   up to FLOW2(20) value.
etc

—&END
```

The storage of the data for, say, the UNIT vector is that UNIT1(30) is allocated the first column of the UNIT(30,50) vector and so on.

The EXECUTIVE routine of the program controls the calling of the data reading routine (DLOAD), the data modification routine (DALTER) and the modules in the correct order of LIST (via MODULE) as shown in flowchart, figure 6.2. Reference to the program in the appendix should provide a clearer view of the system. It has been found that undergraduates can readily understand this system after a relatively short period of study and can write their own subroutines successfully. As with all these types of simulator packages the range of usage is only limited by the number and applicability of the subroutines and progress can always be made as more simulations are attempted and further subroutines become available. New routines should be thoroughly tested before use due to the difficulty in tracing errors caused by slightly incorrect routines. Thus a conventional program can first be written and tested. It should then be tried as a one module plant using the simulator. Trace facilities can be used to print output options.

A number of variations to the above approach were considered when deciding on this approach in PRIMER. They included whether to abolish JUNCTION and permit only one input stream to all routines bar MIXER. This would have permitted a more consistent unit array format but resulted

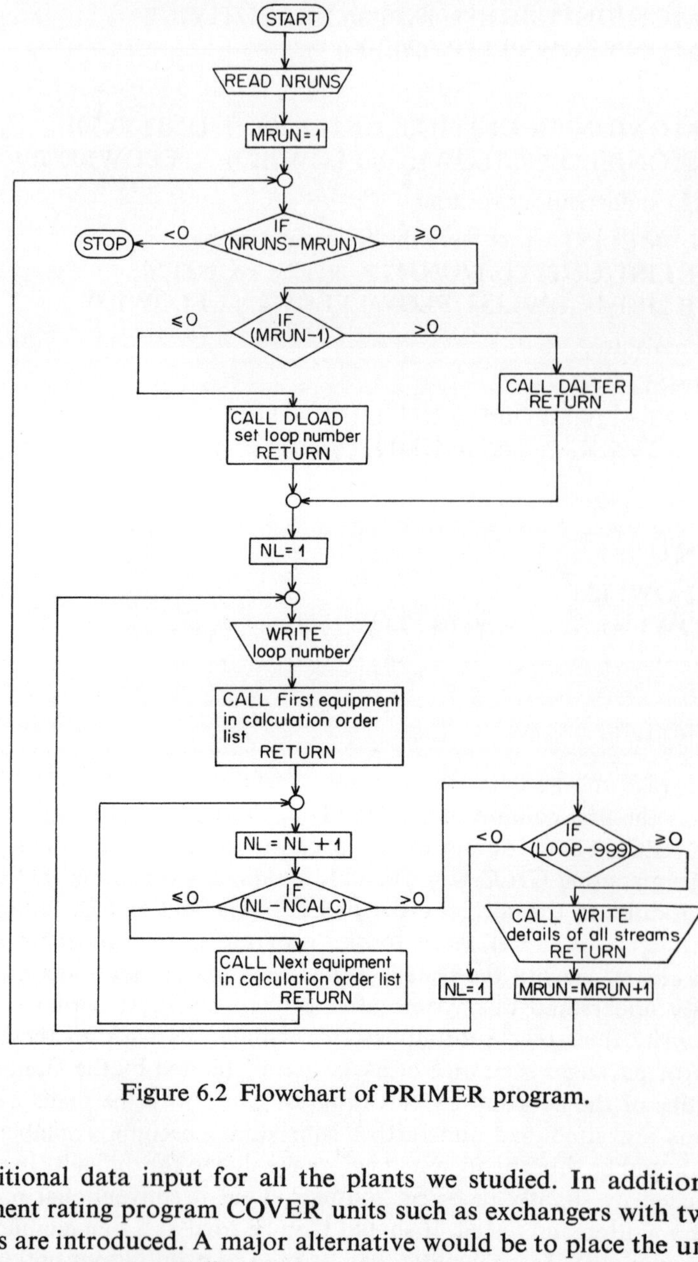

Figure 6.2 Flowchart of PRIMER program.

in additional data input for all the plants we studied. In addition in our equipment rating program COVER units such as exchangers with two input streams are introduced. A major alternative would be to place the unit array format as an argument list in the subroutine name. Thus the subroutine SEPAR 1 could be called as

SUBROUTINE SEPAR 1 (I,J,K,R,S,T,U.)

where the argument list is

I = feed stream number
J = overhead exit stream number
K = bottoms exit stream number
R,S,T,U = fraction components 1–4 in overhead stream.

The expanded version of PRIMER only differ in that an array PROPS (L,I) is introduced where I is the component number and L is a physical property according to an appropriate schedule. The values in this schedule are accessed for each component from a major data bank before using the simulator. Subroutines are provided to compute enthalpy temperatures, etc for the particular mix of components and pressure.

6.4 Dynamic simulation

The process simulators are mainly concerned with steady-state behaviour of chemical engineering systems. For many processes this may be sufficient. However, no process is ever truly in steady-state. Disturbançes occur such as problems in plant operation, change of raw material specification, catalyst vatiation, and for batch processes a time scale must be introduced. Plant performance may be significantly improved by controlling closer to specific tolerance. Examples include optimum reaction temperatures or reducing plant recycle by less impurities in the product from a column. The dynamic information flow diagram of processing units bears close relationship to the instrumented process flow diagram. This increases the library of subroutines for the modular approach, to incorporate aspects such as parabolic liquid flow valves, general well mixed vessel, 3 term controller, etc. The print-out incorporates real time.

DYNSYS (Ponton–Johnson, 1972) is one such modular system, or alternatively one can write one-off programs and incorporate differential equation solving systems such as CSMP (I.B.M., 1967) and MIMIC (Peterson and Sansom). Franks (1972) gives valuable information with respect to this approach.

6.5 Economic simulation

6.5.1 NEW PROJECTS

For new projects it is essential to add cost information to the design model to evaluate capital and operating costs for optimization of process route. At this stage of the process full capacity of plant would normally be used, although one not recommended practice is to optimize at 80% of capacity and assume bottlenecks can be obviated. Clearly it saves the engineer's time if operating and capital costs are evaluated on the simulator. Operating costs can be obtained by generating cooling water, steam, fuel oil, etc in the module and sending information as component streams into a module for their costing. The stream array contains data on the cost/unit of these materials. For optimization of specific process route wages, raw material costs and product revenue is often not required (the same in each case). If required these costs may be evaluated or fed as direct costs into the cost module. As it is desirable to write out a name of the stream other than number and list of components a report writing routine can be added. A yield of product may be estimated and fed as an equipment parameter. Other fixed costs are added following capital estimation when factors are used to assess the costs.

The estimation of capital cost is the greatest problem and accuracy of ± 15 per cent is difficult to achieve. However, as relative costs are compared for process route this tends to narrow this range of error. Delivered equipment costs for use on the simulator are provided either as correlations or as data arrays based on published or past experience. Factors are added to allow for pressure, etc. Materials of construction must be specified as equipment parameters. Other information is normally obtained from the rating module subroutine which must contain sufficient information for the cost routine to function. Thus for a reactor the pressure, approximate wall thickness and volume are specified by the reaction subroutine and material of construction, heating factors such as coil or jacket are specified as equipment parameters. For heat exchangers specify pressure and surface area in EXCHANGER and material of construction in COST EXCHANGER.

The provision of a routine for evaluating installed costs of equipment is more involved. It is necessary for reasonable accuracy to consider each unit separately and this means considering a range of factors as indicated in the program description statements taken from our subroutine INSTALL, table 6.4. Once accurate factors have been obtained it is possible to indicate by a zero or one in the data space which factor to use. Fortunately for optimization purposes the factor required is frequently the same for each unit so once the data has been prepared it is not necessary for it to be changed on each run through the simulator.

For evaluation of capital cost once optimization of route is achieved it may be preferred to run a self-standing cost program. Provision may then be made to incorporate units not included in the simulator (frequently storage tanks) and overwrite delivered equipment costs when obtained as budget quotations from fabricators.

For project evaluation capital cost estimation may be carried out for several plant capacities usually based on a favoured route. Alternatively factorial methods can be used. For a specific capacity of plant it may be desired to evaluate several alternative throughputs to obtain operating costs. To do so requires adjustment of the module subroutines as the objective has changed from design of items to evaluating their performance. Although this is possible it tends to be expensive in computing time and normal estimation is in our opinion preferable.

```
C  DATA INPUT AS ZERO OR ONE
C  FN(1) =   MUCH OF SITE ERECTION INCLUDED IN PURCHASE
C            COST OF EQUIPMENT E.G. LARGE TANKS
C  FN(2) =   AVERAGE ERECTION
C  FN(3) =   EQUIPMENT INVOLVING SOME SITE FABRICATION
C            E.G. LARGE PUMPS REQUIRING LINING UP,
C            SERPENTINE COOLERS
C  FN(4) =   EQUIPMENT REQUIRING MUCH SITE FABRICATION
C            OR FITTING E.G. LARGE DISTILLATION COLUMNS,
C            FURNACES
C            AT DISCRETION OF ESTIMATOR, INTERPOLATION
C            MAY BE MADE
C  FN(5) =   ZERO
C  FACTORS FOR PIPING, DUCTING AND CHUTES INCLUDING
C            ERECTION
C
C  FN(6) =   DUCTING AND CHUTES
C  FN(7) =   SMALL BORE PIPING OR SERVICE PIPING ONLY
C  FN(8) =   AVERAGE BORE PIPING AND SERVICE PIPING (E.G.
C            PREDOMINANTLY LIQUID PIPING)
C  FN(9) =   LARGE BORE PIPING AND SERVICE PIPING (E.G.
C            PREDOMINANTLY GAS AND VAPOUR PIPING)
C  OR        AVERAGE BORE PIPING WITH COMPLEX SYSTEM
C            E.G. MUCH MANIFOLDING, RECIRCULATION ETC)
C  FN(10) =  LARGE BORE PIPING COMPLEX SYSTEM (E.G. MUCH
C            MANIFOLDING, RECIRCULATION ETC)
C  FN(11) =  MULTIPLY PIPING FACTORS BY 1.3 FOR SPECIAL
C            PIPE MATERIALS OR STEAM TRACING
C  FN(12) TO FN(15) EQUAL ZERO
C  FACTORS FOR INSTRUMENTS
C
C  FN(16) =  LOCAL INSTRUMENTS ONLY
C  FN(17) =  1 CONTROLLER AND INSTRUMENTS
C  FN(18) =  2 CONTROLLERS AND INSTRUMENTS
C  FN(19) =  3 OR MORE CONTROLLERS AND INSTRUMENTS
C  FN(20) =  ZERO
C  FACTORS FOR ELECTRICAL
C
C  FN(21) =  LIGHTING ONLY
C  FN(22) =  LIGHTING AND POWER FOR ANCILLARY DRIVES
C            (E.G. CONVEYORS, STIRRED VESSELS, AIR COOLERS,
C            ETC)
C  FN(23) =  LIGHTING AND POWER EXCLUDING TRANSFORMERS
C            AND SWITCHGEAR (I.E. THIS EQUIPMENT OFF SITE)
C              FOR MACHINE MAIN DRIVES (E.G. PUMPS,
C              COMPRESSORS, CRUSHERS, ETC)
C  FN(24) =  LIGHTING AND POWER INCLUDING TRANSFORMERS
C            AND SWITCHGEAR
C              FOR MACHINES MAIN DRIVES (E.G. PUMPS,
C              COMPRESSORS, CRUSHERS, ETC)
C  FN(25) =  ZERO
```

Table 6.4 Factors for main plant items erection.

```
C  FACTORS FOR CIVIL
C
C  FN(26)=  AVERAGE CIVIL WORK, INCLUDING PLANT AND
C           STRUCTURE FOUNDATIONS, FLOORS AND SERVICES
C  FN(27)=  ABOVE AVERAGE CIVIL WORK, COMPLICATED
C           MACHINE BLOCKS, SPECIAL FLOOR PROTECTION,
C           ELEVATOR PITS IN FLOORS, CONSIDERABLE
C           SERVICES
C  FN(28) TO FN(29) EQUAL ZERO
C  FN(30)=  MULTIPLY CIVIL FACTOR BY 1.3 TO ALLOW FOR
C           PILING PLANT AND STRUCTURE FOUNDATIONS
C  FACTORS FOR STRUCTURES AND BUILDINGS
C
C  FN(31)=  NEGLIGIBLE STRUCTURAL WORK AND BUILDINGS
C  FN(32)=  OPEN AIR PLANT AT GROUND LEVEL WITH SOME
C           PIPEBRIDGES AND MINOR BUILDINGS
C  FN(33)=  OPEN AIR PLANT WITHIN A STRUCTURE
C  FN(34)=  PLANT IN A SIMPLE COVERED BUILDING
C  FN(35)=  PLANT IN AN ELABORATE BUILDING OR ON A
C           MAJOR STRUCTURE WITHIN A BUILDING
C  FACTORS FOR LAGGING
C
C  FN(36)=  LAGGING FOR SERVICE PIPES ONLY
C  FN(37)=  AVERAGE AMOUNT OF HOT LAGGING ON PIPES
C           AND VESSELS
C  FN(38)=  ABOVE AVERAGE AMOUNT OF HOT LAGGING ON
C           PIPES AND VESSELS
C  FN(39)=  COLD LAGGING ON PIPES AND VESSELS
C  FN(40)=  ZERO
C  ONLY ROUTINE WHEN DELIVERED CAPITAL COST IS ≥£20,000,
C  IS INCLUDED
   COMMON EN(40),FN(40),NT,DEC,DIC,INST,XNT(20),DELC(20),K
   IF(DEC.GE.20000.0)GO TO 201
   IF(DEC.GE.5000.0)GO TO 202
   IF(DEC.GE.3000.0)GO TO 203
   IF(DEC.GE.1500.0)GO TO 204
   IF(DEC.GE.1000.0)GO TO 205
   IF(DEC.GE.750.0)GO TO 206
   GO TO 207
201 FEREC=FN(1)*0.01+FN(2)*0.04+FN(3)*0.06+FN(4)*0.24
   FPIP=FN(6)*0.02+FN(7)*0.05+FN(8)*0.13+FN(9)*0.16+FN(10)*0.20
   IF(FN(11).EQ.1.0.)FPIP=FPIP*1.3
   FINST=FN(16)*0.02+FN(17)*0.07+FN(18)*0.10+FN(19)*0.15
   FELEC=FN(21)*0.02+FN(22)*0.08+FN(23)*0.10+FN(24)*0.15
   FCIV=FN(26)*0.06+FN(27)*0.12
   IF(FN(30).EQ.1.0)FCIV=FCIV*1.3
   FBUILD=FN(31)*0.01+FN(32)*0.05+FN(33)*0.11+FN(34)*0.15+FN
  1 (35)*0.25
   FLAG=FN(36)*0.01+FN(37)*0.02+FN(38)*0.03+FN(39)*0.05
   DIC=DEC+(DEC*(FEREC+FPIP+FINST+FELEC+FCIV+FBUILD
  1 +FLAG))
   RETURN
```

Table 6.4 Factors for main plant items erection—*continued*

For authorization of project expenditure several cases must be examined and the appropriate economic criteria used to determine the best case. Such criteria as N.P.V., Venture Worth, D.C.F., involve estimation of cash flows allowing for time value of money. Self-standing programs are readily devised for this purpose (Allen, 1972).

6.5.2 EXISTING PLANT

Johnson (1971) has applied the modular approach to planning aspects relevant to the operation of chemical plants. Examples of such activities are production planning, determining raw material requirements, cost analysis and profit planning. We have produced a version of PRIMER expressly for this purpose. This has as its basis the observation that *in corporate business simulation information flow is the dominant consideration*. Thus it may be decided to have the process products as main input to the model. Such a demand orientated model outputs the intermediate product amounts and raw material requirements. Alternatively, if raw material availability is the planning criterion material flow tends to be followed and intermediate raw materials and products are the output. Between the two methods recycle problems can usually be avoided. If not, either a Wegstein or simultaneous method may be used.

Consider the following problem. Aniline is produced from nitrobenzene which in turn consumes sulphuric acid. Spent raw materials are recycled back to source as shown in figure 6.3(a). Introducing a demand orientated model results in reversal of product flow and output of raw materials and intermediate products. As shown in figure 6.3(b) the resulting structure is acyclic.

The incorporation of a cost/unit in the FLOW (4,*I*) location enables the cash flow in any stream to be computed once the mass flow is evaluated.

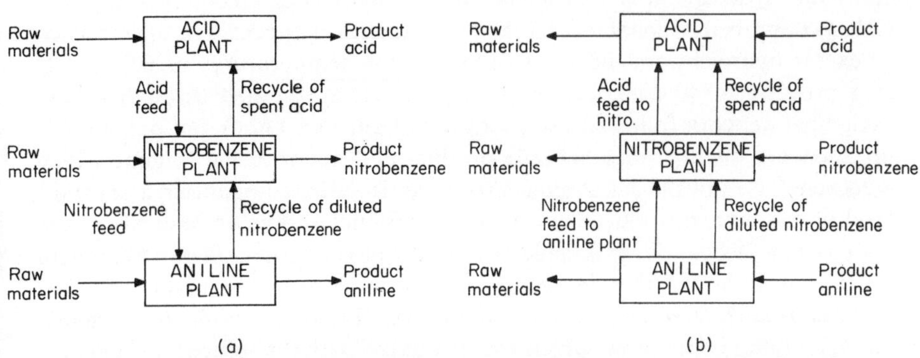

Figure 6.3 Modifying calculation structure in demand orientated model. (a) Actual material flow exhibits recycle structure. (b) Information flow exhibits acyclic structure.

For sound report writing virtually all streams are stored. The input can be large and NAMELIST unformatted input is essential for rapid use. Additional modules required include a DIFFERENCE routine which subtracts one cash flow from another. Negative flow may also be obtained if required by negating the cost/unit. The PANDL routine carries out cash summation and writes the report in the form of a profit and loss account. A CONSTRAINT routine permits ensuring that maximum plant capacity is considered.

Burgess and Wells (1973) have reported on simulation of small chemical works in which a block corresponds to each plant on the works. Factors are fed in to indicate the split of the total mass flow at each unit and the variable costs which arise. Product and variable operating costs are the output to the report writing module. Direct costs are also input to this unit. Similar results could be achieved using a linear programming system. However, for this type of scale and use we find that the modular approach reflects better the attitudes of ordinary managements and is extremely flexible. It is also easier to establish the initial model. Clearly where more conflicting parameters exist a linear programming model has its advantages in that it will perform an optimization exercise. A further approach under development is to use the modular method to generate part of the L.P. matrix (rather as SYMBOL produces process mass balances) and then add in any constraints. The resulting matrix may be solved in the conventional manner.

6.6 Case study

6.6.1 PROCESS DESCRIPTION

In order to illustrate the principles of computer aided design a plant involving several recycles has been devised, figure 6.4. The process flow diagram rather than a block diagram is shown for clarity of presentation. The process concerns the hydrocracking of toluene to benzene $C_6H_5.CH_3+H_2 \rightarrow C_6H_4+CH_4$ as described in chapter 5.1.1. Briefly toluene is vaporized in the presence of excess hydrogen and heated to the reaction temperatures of 600–800°C at a pressure of 40 bar. The reaction products are cooled through a heat exchanger network followed by separation of gas in a knock-out pot. Liquid products are distilled into benzene product and unconverted toluene which is recycled. Part of the gas stream is recycled. In order to maintain a constant level of methane, introduced as a reaction product and as an impurity in the hydrogen, a fuel gas purge is made prior to the introduction of fresh hydrogen.

It is emphasized that in reality such a process would involve further operations such as provision to accommodate (1) side reactions to diphenyl, etc, (2) sulphur compounds which are introduced with the toluene and necessitate a clean-up stream on the gas recycle, (3) a stabilizer to remove trace gas in the liquid streams. The system is simplified solely in order to permit ready comprehension of the computation stages.

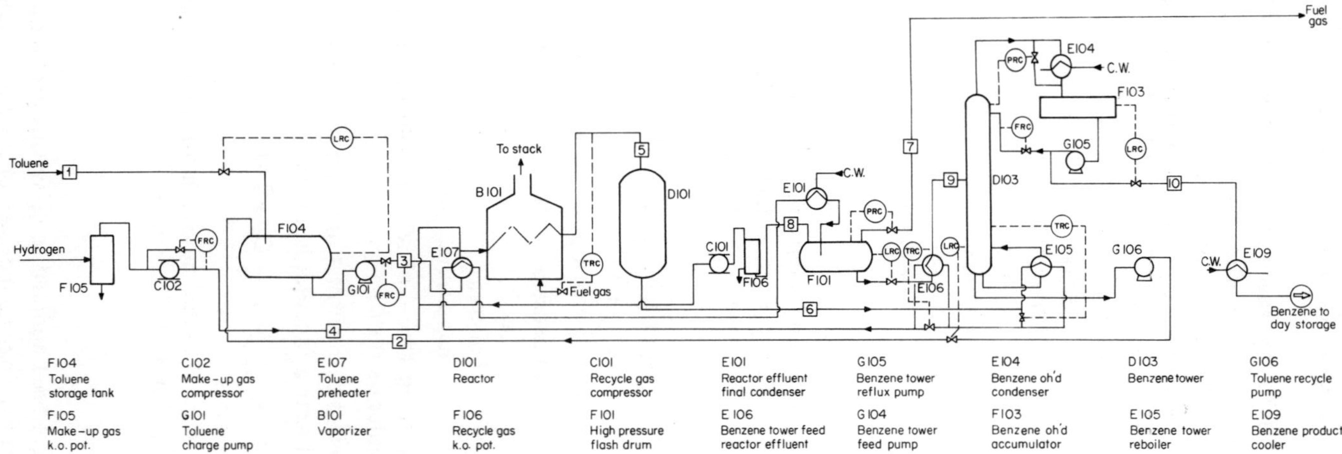

Figure 6.4 Process flow diagram toluene hydrocracker.

6.6.2 DESIGN CRITERIA

(a) The process is to be designed to convert 10 Mg/h of toluene to benzene.

(b) The molar ratio of hydrogen : toluene should be 5:1 at the inlet to the reactors. Operating pressure of reactors to be 40 bar.

(c) Information on the reactor system indicated that for a gas flow of 20 000 m^3/h (S.T.P.) reactor costs, conversion and operating temperatures are

Conversion (%)	Installed cost (£)	Reaction temperature (°C)
70	25 000	600
75	30 000	675
90	50 000	750

(d) Operating hours: 8 000 h/annum
Hydrogen feed: 70% wt hydrogen
30% wt methane

(e) Operating costs:

electricity	£5/MWh
cooling water	£0.00125/Mg
fuel gas	£0.00025/MJ
hydrogen	£45/Mg

(f) Capital costs:

Vaporizer £0.75/MJh

All other units from Capital Cost Estimation, *Inst. Chem. Eng.*, 1969.

(g) Heat exchanger overall heat transfer coefficients may be assumed as

condensers	400 W/m^2 degK
gas/gas	100 W/m^2 degK
reboilers:	
max flux	30 000 W/m^2

Assume approach to temperature of reactor effluent before final condenser 140°C, and after condenser 30°C.

(h) Assume installed cost of total unit is three times delivered costs.

(i) The design objective function is encoded as

$$U_{min} = C_V + 0.4I$$

where I is installed cost
C_V is variable costs excluding overheads and maintanance etc
U is objective function £/annum.

6.6.3 STRATEGY

Several strategies have been adopted to solve this problem according to our availability of programs. They include use of process simulators as follows:

for mass balance only, SYMBOL, Nagiev (subroutine for solving simultaneous equations), GEMCS, PRIMER: for heat and mass balances, CHESS: for equipment sizing and heat/mass balances, CHESS. Further programs were used for sizing distillation columns by Fenske–Underwood, heat exchangers (assuming overall heat transfer coefficients), vaporizer, pumps, compressors. Better programs available for distillation and exchanger design were not used in order to reduce computing time. An algebraic approach was used as an alternative to the simulators for formulating a one-off program. Additional programs were written as required.

The method of attack is shown in figure 6.5. In each case an information flow diagram is constructed based on the process flow diagram supplied. The appropriate simulator/sizings are used to supply data for costing which is carried out manually. The system variables are adjusted to obtain the most economic solution of the system.

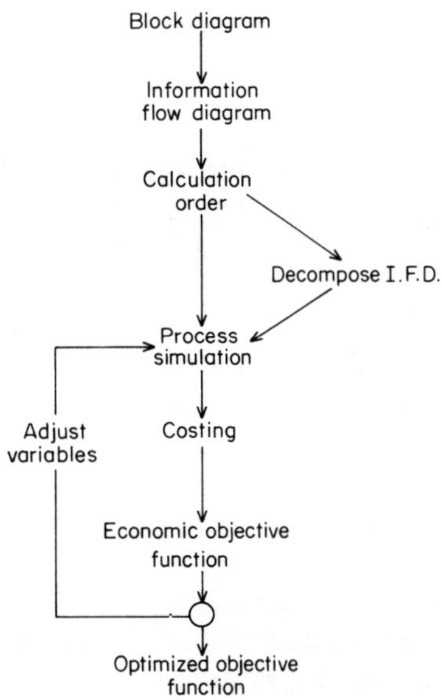

Figure 6.5 Basic strategy.

1 *Information flow diagrams*

Three information flow diagrams (I.F.D.) were constructed to permit different methods of solving the problem. Details of streams and modules are given in table 6.5 and apply to figures 6.6, 6.7 and 6.8. The numbering indicated on the diagrams is for consistency in this presentation, different figures being used in practice.

The main emphasis in these diagrams is on efficient partitioning. The first diagram simulates the flowsheet by substituting computation blocks to replace units, figure 6.6. Although the problem apparently contains only two recycle sets these are nested in such a difficult manner as to mean that calculation was carried out in the sequence of modules (1 2 3 4 5 6 7 8 9 10 11), guessing streams 7, 10 and 16 and converging streams 7 and 8.

The second I.F.D., figure 6.7, contains an attempt to break the recycle loop by carrying out a mass balance followed by a heat balance along an acyclic path, i.e. modules (1 3 7 11 8 9 10), guessing 10, 16, converging on stream 9 followed by 4, 5, 2.

The third I.F.D., figure 6.8, shows application of a program ordering routine to identify the cut streams and, if values of these streams are known, feeding such information into the system. This is a feature more used in design than for simulation of an existing process. For economic evaluation there is a greater tendency to think about what do we know and less concern about modelling exactly the condition of each stream. In this case we know as design criteria the temperature of streams 4 and 9 but not the mass flow. However the latter may readily be obtained by using a simplified mass balancing routine. The major difference between figure 6.7 and 6.8 relates to their use in the techniques which follow. These are:

1 (a) Simultaneous solution of the mass balance, using SYMBOL.
(b) Construction of a heat balancing routine which sized heat exchangers and vaporizer, using acyclic structure as indicated in figure 6.7. A reflux ratio was assumed for the columns thus permitting rough estimation of the reboiler duty.
(c) Sizing the column and prime movers separately.

2 (a) GEMCS or PRIMER for mass balances and data as to temperature and pressure at any point, assuming a reflux ratio for the column.
(b) Sizing units separately based on data provided from the simulator.

3 (a) CHESS in unit modules as provided, without modification of I.F.D. structure, that is as in figure 6.8.

4 (a) Simultaneous solution of the mass balance.
(b) CHESS on using above information for rapid convergence and sizing equipment with acyclic I.F.D. as in figure 6.7.

5 Algebraic formulation of the optimization problem and constructing an overall program.

2 *Optimization*

The cases were attempted by final year students as part of their design problem. The results give an insight into the difficulties which arise from the inexperienced user.

Optimization of the process involves varying the % conversion, the % benzene in the recycle of toluene, the reaction temperature and the quantity of hydrogen in the make-up stream. The latter has the greatest effect on the process affecting raw material costs, fuel gas credit and the size of the recycle

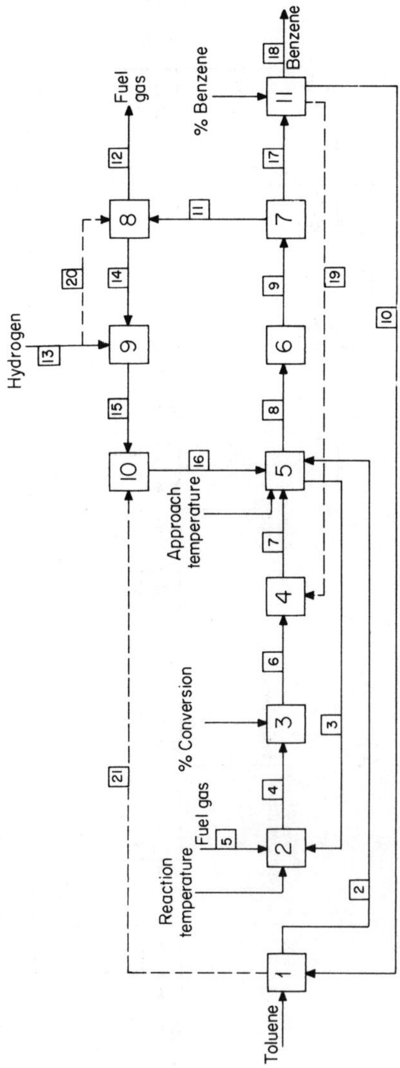

Figure 6.6 Information flow diagram before modification.

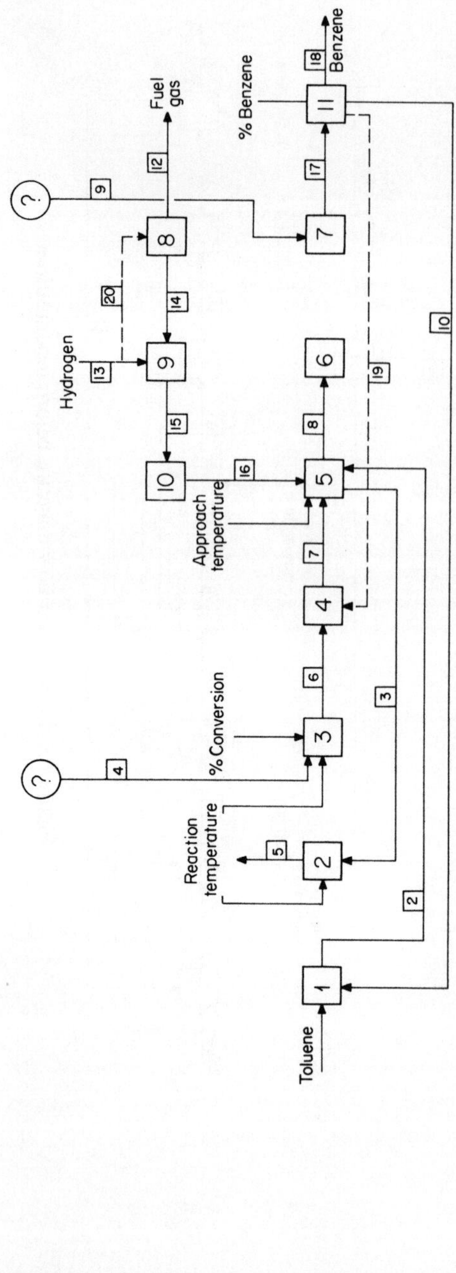

Figure 6.7 Information flow diagram after decomposition to acyclic structure.

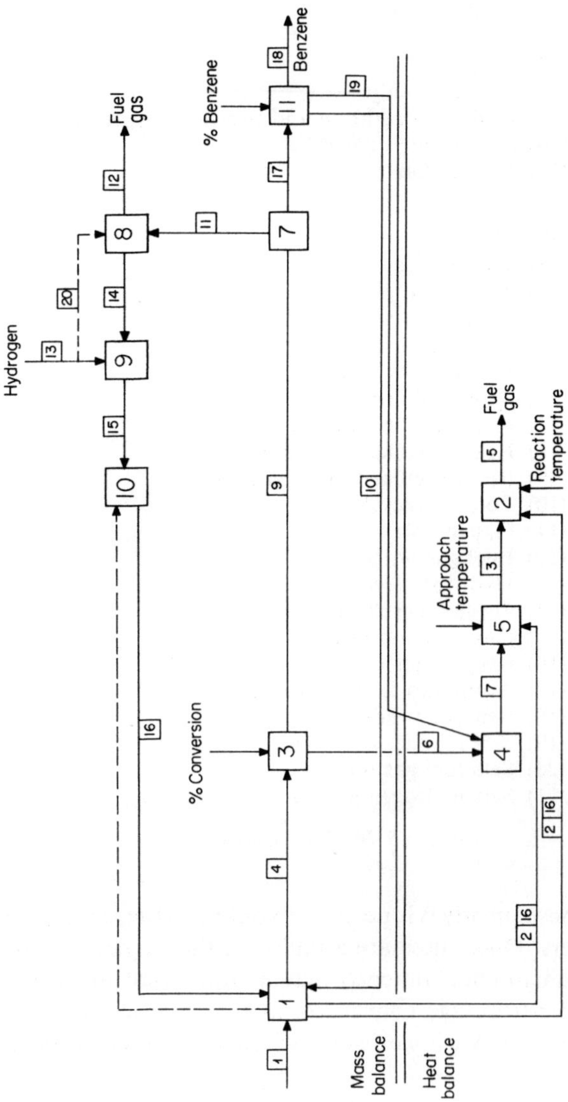

Figure 6.8 Information flow diagram after modification of heat balance.

Modules

(1) Toluene feed (pump and junction)
(2) Vaporizer
(3) Reactor
(4) Reboilers
(5) Heat exchange
(6) Condenser
(7) Separator
(8) Fuel gas purge
(9) Hydrogen addition and recycle compressor
(10) Set toluene : gas ratio
(11) Distillation

Streams

(1) Toluene feed
(2) Toluene to preheat
(3) Toluene to vaporizer
(4) Toluene to reactor
(5) Fuel gas consumed
(6) Reactor effluent to reboiler
(7) Reactor effluent from reboiler
(8) Reactor effluent to condenser
(9) Reactor effluent to separator
(10) Toluene recycle
(11) Gas recycle
(12) Fuel gas purge
(13) Hydrogen feed
(14) Purged recycle gas
(15) Gas to compressors
(16) Recycle gas to preheat
(17) Aromatics feed to distillation
(18) Benzene product
(19) Sets reboiler heat load
(20) Sets fuel gas offtake
(21) Sets hydrogen gas : aromatics ratio

Table 6.5 Modules and streams.

compressors. Consequently it is possible to optimize these parameters without varying the others. Once these are established the % benzene in the recycle which has only small effect on costs may be optimized independently of the temperature and percentage conversion in the reactors which are the final items to be optimized. Variables were adjusted as indicated in table 6.6.

Hydrogen feed Mg/h 0.1, 0.2, 0.4, 0.6, 0.8, 1.0.	
Benzene in recycle toluene %wt	2.5, 4, 7.5, 10
Conversion %	70, 75, 80, 85, 90
Temperature °C	1 100, 1 200, 1 300

Table 6.6 Variables for optimization.

3 *Results and discussion*

The results of the optimization exercise are shown in table 6.7 and show how the costs were adjusted by the variables. A major criticism is that none of the simulators used contained a costing routine neither did they provide adequate information about equipment to obtain costs directly from graphs, etc.

For this particular problem the algebraic solution of the mass balancing problem using simultaneous solution was trivial and is reported in chapter 5. It was the quickest means of carrying out all mass balances. The use of the acyclic structure of the information flow diagram permitted ready construction of a three-stage heat balancing program. Similarly a modified version of SYMBOL for simultaneous solution of the mass balance proved the easiest simulator to use.

The main problem with the simultaneous solutions was the absence of efficient routines to alter data once a set of results had been obtained. This was a serious disadvantage for this particular problem. Both the GEMCS system and PRIMER required the adaptation of two modules for mass balancing. Additionally, any design change made in temperature and pressure was indicated. Thus complete stream information was made available for sizing. Efficient ALTER routines permitted rapid assembly of each set of data. The Wegstein convergence routine proved vital. It was generally felt that the amount of effort made to achieve values would be more worth while if some corresponding ratings had been obtained.

The simulator CHESS was misused by the users in group 3. Having a simulator capable of rating equipment meant an increase in the number of recycle loops to include exchangers and a tendency for the routines not to converge in the time available. Far more successful was the use of CHESS after modifying the IFD structure and using the modules for sizing equipment following either convergence of mass balances or input of accurate balance figures from PRIMER. The physical properties package of this simulator was invaluable, not only for this group but as a ready source of data. Again several additional routines were required and the students found this a difficult task. The size of this simulator tended to overwhelm the students as did its complexity. Run failures did occur, often due to trivial errors, and sometimes as a result of not fully understanding the machinations of the simulator. Certainly a greater specialist knowledge is demanded by these larger programs. The problem of run failure is also more severe as such a large core requirement as CHESS may mean allocation of a job as say ‘special’ priority, thus giving no chance of achieving several runs in a day. Five trivial errors can thus mean a loss of one week on a project. On the other hand, a success does mean more rapid evaluation than in other methods.

Algebraic solution of the problem proved the simplest technique as might be expected for such a trivial problem. Following debugging of the individual sections of the plant the programs were combined. Each variable was adjusted

CASE	VARIABLES			CAPITAL COSTS				OBJECTIVE FUNCTIONS					
	% benzene in recycle	% conversion of toluene	% weight of hydrogen in fuel gas	Prime-movers £	Reactor £	Heaters £	Column £	Hydrogen less fuel gas credit	Prime-movers £ p.a.	Reactor £ p.a.	Heaters £ p.a.	Column £ p.a.	Total £ p.a.
1	2.5	70	19.16	40 800	29 600	65 200	41 500	199 700	27 900	11 800	77 900	18 800	336 100
2	2.5	70	14.78	40 000	31 600	70 000	41 500	154 700	27 000	12 600	83 900	18 800	297 000
3	2.5	70	9.55	41 300	35 800	76 300	41 500	110 000	26 200	14 300	91 200	18 800	260 500
4	2.5	70	3.43	48 100	53 700	109 400	41 500	65 100	31 800	21 500	131 000	18 800	268 200
5	2.5	70	0.98	180 000	102 500	236 600	41 500	46 000	155 000	41 000	289 000	18 800	549 800
6	4.0	70	9.55	41 000	36 000	75 400	39 800	110 000	26 000	144 000	90 100	18 100	258 600
7	7.5	70	9.55	41 200	36 200	75 900	37 400	110 000	26 200	14 500	90 600	17 100	258 400
8	10.0	70	9.55	41 400	36 500	76 200	36 200	110 000	26 500	14 600	91 000	16 600	258 700
9	7.5	75	9.55	40 000	41 400	71 100	37 200	110 000	25 600	16 600	84 300	16 700	253 200
10	7.5	85	9.55	35 500	52 300	63 300	37 200	110 000	24 900	20 900	74 700	16 700	247 200
11	7.5	85	14.78	37 500	49 600	57 600	37 200	155 000	24 800	19 800	68 000	16 700	284 300
12	7.5	85	3.43	42 700	84 200	91 200	37 200	65 100	27 700	33 700	107 800	16 700	251 000
13	**7.5**	**90**	**9.55**	**34 100**	**60 100**	**60 500**	**34 300**	**110 000**	**22 400**	**24 000**	**70 900**	**15 500**	**242 800**
14	7.5	90	3.43	40 900	90 200	86 900	34 300	65 100	26 400	36 100	101 900	15 500	245 000
15	7.5	90	14.78	36 200	53 000	55 000	34 300	154 700	24 000	21 200	64 800	15 500	280 200
16	10.0	90	9.55	34 200	60 400	60 600	33 300	110 000	22 600	24 200	71 200	15 000	243 000
17	10.0	90	3.43	41 000	90 400	87 000	33 300	65 100	26 600	36 200	102 400	15 000	245 300
18	4.0	90	9.55	33 900	60 000	60 400	36 200	110 000	22 300	24 000	71 000	16 100	243 400

Narrowing the search further indicated an optimum value at 6% benzene in recycle, 7% hydrogen in fuel gas, 90% conversion of toluene.

Table 6.7 Economic analysis of toluene hydrocracker.

until all cases had been studied. The result of completely integrating the problem was the most accurate solution. It should be noted that some existing self-standing programs were integrated in this package.

The self-standing programs for distillation and heat exchangers proved useful as they provided cost information.

This exercise may be completed using the Hooke and Jeeves package of chapter 7 to optimize the heat exchanger/vaporizer network. The use of an optimization routine to automate the search for the best overall process solution was not possible on our facilities. If used injudiciously it could mean varying each variable independently and over 400 designs of the system, instead of the 18 actual ones attempted. It might be added that time would have been greatly reduced had the costing been automated and the longer runs been possible on the computer. An interactive facility would only marginally have affected the time involved.

The conclusions from this study and similar reports support the view that *for design a study of information flow is vital rather than blind use of simulation.* Also rapid convergence of recycle loops is desirable and hence simultaneous solution or efficient convergence promotion. This must be combined with an efficient ALTER routine to permit processing several sets of data once the program has been compiled. To reduce the fag of card punching, free format is desirable.

The availability of simulators determines choice. Clearly we want an efficient simulator probably available in segments as required, per use, and capable of rating equipment, with costing subroutines, not occupying too great a core size for the local machine and of sufficient simplicity to write additional routines with ease. Failing this there is merit in carrying out preliminary mass balances only on a smaller simulator or one-off derived program, and using such information as the feed to either self-standing or smaller simulation programs, capable of carrying out some of the more common unit operations and preferably producing equipment rating and costing in a sequence predetermined as acyclic.

Notation	Definition	Section
C_v	variable cost	6.5.2
I	installed cost	6.5.2
I	module number	6.2
I	stream number	6.2
J	first stream array element	6.2
JJ	length of stream array	6.2
N	first module array element	6.2
U	design objective function	6.5.2
Subscripts		
$_{min}$	minimum	6.5.2

7 General Optimization Problems

Optimization problems frequently necessitate use of the computer because of the vast number of alternative cases and variables which require investigation. Obviously the strategy to be adopted depends on a full knowledge of optimization theory which is not in the remit of this book. A number of problems are indicated here with emphasis solely on use and development of programs. Furthermore, for these examples we do not intend to discuss or detail the formulation of the correct objective function as was developed previously for these examples (see Wells, 1973). The reader needs to be aware of these methods so that he can apply them as they arise. Here emphasis is on final optimization of the units and system of which they are part. A few simple examples are considered before the larger systems. This permits covering a slightly wider range of optimization problems within the same chapter.

7.1 Search over a single variable using discrete sizes

Hot service insulation of pressure vessels

The hot service insulation of large vessels may be approximated as a flat surface. Assuming fluid film coefficients are negligible the heat loss, q, is given by

$$q = \frac{kA\Delta t}{x}$$

where k = conductivity
A = area of surface
Δt = temperature difference
x = thickness.

Assuming 8 600 h operating time per year and considering unit area, then:

$$\text{Value of heat loss} = cq\ 8\,600$$

c = cost of steam/unit of heat

$$\text{The investment costs} = (I \times d) + i_m(1-g)/(1-t)$$

I is investment
d is depreciation

i_m is minimum acceptable return
g is investment grant
t is taxation.

The economic objective function to be minimized is taken as simple Venture Profit (Wells, 1971)

$$U = \frac{8\,600\,k\,c\,t}{x} + (I \times d) + i_m(1-g)/(1-t)$$

Assuming t independent of x we may differentiate and set the derivative to zero.

Then
$$x^2 = \frac{8\,600 kc\Delta t}{(I \times d) + i_m(1-g)/(1-t)}$$

It is a simple matter to evaluate the insulation thickness over a range of internal temperatures for discrete sizes of insulation up to 80 mm. This will produce, in tabular form, recommended insulation thickness schedules which may be used for quick reference on the project. The procedure in slightly refined form then permits of a more accurate solution to the problem than using the same rule of thumb independent of changing economic climate.

Often similar procedures exist for cases where integration is not readily possible, one of these being to search the appropriate equation for a solution.

For equipment obtained in a limited number of discrete sizes the procedure can be to substitute the sizes into the appropriate equations and compare all answers for the solution. If the function is unimodal savings can be made by stopping the substitution once a minimum value is achieved. A suitable exercise is the selection of optimum pipe diameters (Wells and Bush, 1971).

7.2 Unimodal search

If the function is unimodal a solution to the required accuracy may be achieved by searching over the region using the values obtained from the initial search points for direction of the search. Such a method is that of the Golden Section which may be used to evaluate the Underwood parameter, Θ, mentioned earlier (4.4.5). The equation is

$$\sum \frac{\alpha_i x_{iF}}{\alpha_i - \Theta} - (1+q) = 0$$

where α_i = relative volatility of ith component
x_{iF} = mol fraction of component in feed
q = state of feed.

Ignoring the sign of the function, search for the value of Θ within the relative volatilities of the key components that yields the minimum value of the function.

```
      MASTER SECTION
      COMMON/GOLD1/M,FQ,RV(10),XF(10)
      NAMELIST/ALIST/A,B,M,FQ,XF,RV
      READ(2,ALIST)
      CALL GOLDEN(A,B)
      CONTINUE
      CALL EXIT
      END

      SUBROUTINE GOLDEN(A,B)
  100 GL=B-A
      R=A+0.382*GL
      CALL FUNC(R,X)
      CONTINUE
      S=B-0.382*GL
      CALL FUNC(S,Y)
      CONTINUE
      D=Y-X
      C=S-R
      IF(C-0.001)104,104,101
  101 IF(D)103,103,102
  102 B=S
      GO TO 100
  103 A=R
      GO TO 100
  104 IF(D)106,106,105
  105 VAR=R
      GO TO 107
  106 VAR=S
  107 CALL REPORT(VAR)
      RETURN
      END

      SUBROUTINE FUNC(T,UA)
      COMMON/GOLD1/M,FQ,RV(10),XF(10)
      DIMENSION ULA(10)
      TOTAL=0.0
      DO 10 I=1,M
      ULA(I)=(RV(I)*XF(I))/(RV(I)-T)
   10 TOTAL=TOTAL+ULA(I)
      URA=1.0-FQ
      UA=ABS(TOTAL-URA)
      RETURN
      END

      SUBROUTINE REPORT(VAR)
      WRITE(3,100)VAR
  100 FORMAT(1H0,'UNDERWOOD PARAMETER=',F10.4)
      RETURN
      END
```

Program 7.1 Search by the Golden Section for Underwood's Θ.

Program 7.1 has been written in sections to keep the function to be evaluated and the search method GOLDEN in seperate segments. In more complex routines this is a useful ploy.

For the two-component system where $\alpha_1 = 5$, $\alpha_2 = 1$, $x_{1F} = 0.6$, $x_{2F} = 0.4$, $q = 0.5$, the function was evaluated as given in table 7.1. The value of Θ is isolated in 12 calculations to the region $1.86 \leqslant \Theta \leqslant 1.80$. Note for instance that if an inexperienced programmer had simply upgraded Θ from Θ_0 by say

$$\Theta_{i+1} = \Theta_i + 0.01$$

this would involve 86 calculations to achieve this degree of accuracy. A useful exercise is to try other methods described in chapter 3 and compare the number of cycles required.

Cycle K	Length L_K	Values of Θ_1	Θ_2	Function $f(\Theta)_1$	Value $f(\Theta)_2$	Region containing
1	4.000	2.528	3.472	0.452	1.301	$1 \leqslant \Theta \leqslant 5$
2	2.472	1.944	2.528	0.058	0.452	$1 \leqslant \Theta \leqslant 3.472$
3	1.528	1.584	1.944	0.307	0.058	$1 \leqslant \Theta \leqslant 2.528$
4	0.944	1.944	2.167	0.058	0.216	$1.584 \leqslant \Theta \leqslant 2.528$
5	0.583	1.807	1.944	0.056	0.057	$1.584 \leqslant \Theta \leqslant 2.167$
6	0.360	1.722	1.807	0.139	0.056	$1.584 \leqslant \Theta \leqslant 1.944$
7	0.222	1.807	1.861	0.056	0.009	$1.722 \leqslant \Theta \leqslant 1.944$
8	0.137	1.861	1.892	0.009	0.017	$1.807 \leqslant \Theta \leqslant 1.944$
9	0.085	1.839	1.861	0.028	0.009	$1.807 \leqslant \Theta \leqslant 1.892$
10	0.053	1.860	1.872	0.009	0.001	$1.839 \leqslant \Theta \leqslant 1.892$
11	0.032	1.872	1.880	0.001	0.007	$1.860 \leqslant \Theta \leqslant 1.892$
12	0.020	1.868	1.872	—	—	$1.860 \leqslant \Theta \leqslant 1.880$

Table 7.1 Underwood's Θ by the Golden Section.

7.3 Logical search

If two or more variables require investigation a search procedure is required which prevents coming to an early conclusion that the first minimum achieved is the optimum. At the same time to search every alternative at this detailed design stage is time consuming and hence expensive.

The most convenient procedure for handling problems of this nature is to have a pack of standard search programs, so eliminating the concept of each design as a one-off job, i.e. provide subroutines for linear programming, Hooke and Jeeves direct search, successive linear approximation, random search, simplex direct search, Davidon, Fletcher, Powell gradient search. Siddall (1970) has developed one such pack, OPTISEP, and we will illustrate the Hooke and Jeeves search from this manual. The subroutines for this purpose as provided in OPTISEP are:

(a) SEEK 1 a master executive
(b) SEARCH Hooke and Jeeves algorithm
(c) SHOT shotgun type pattern change
(d) FRANDN random number generator for SHOT
(e) UREAL defines function being optimized
(f) ANSWER output routine
(g) CONST defines constraints
(h) OPTIMF 1 tests violation of constraint and acts accordingly.

The user has to write two of these routines, UREAL and CONST.

Subroutine UREAL defines the function that the complete program is optimizing, i.e. $U = U(X_1, X_2 \ldots X_N)$. In its simplest form UREAL is written

```
SUBROUTINE UREAL (X, U)
DIMENSION X(N)
U = X(1)+X(2)+ ... +X(N)
RETURN
END.
```

X may be subscripted from 1 to N depending upon the number of variables involved in the optimization function.

Preliminary calculation of the variables may take place in this routine, provided that the final function to be optimized is equated to U. It should also be noted that the program only recognizes the variables $X_1, X_2, \ldots X_N$ and hence each variable incorporated in UREAL must be assigned one of these values.

CONST is another user written routine. It defines the inequality constraints imposed on the problem being solved. For instance, it is usual to look for an optimum between certain limits say an upper limit of H and a lower limit of L. The program needs the information that a solution is required greater than or equal to L and less than or equal to H.

CONST is written in the following form:

```
SUBROUTINE CONST (X, NCONS, PHI)
DIMENSION X(N), PHI (NCONS)
PHI(1) = arithmetic function
PHI(2) = arithmetic function
...
...
PHI(NCONS) = arithmetic function
RETURN
END
```

The routine calculates the inequality constraints at a point

$$\phi_K = \phi_K(X_1, X_2, \ldots X_n) \quad K = 1,p$$

where $\phi_K \geqslant 0$ at a feasible point.

Various functions are defined in addition to the above. These are associated with print parameters, loops, step size, bounds and dimensioning.

Calling routine

Before the program will function a calling program which acts as the master routine, must be written. In this simplest form it is as follows:

(a) Dimension all subscripted variables.
(b) Define all input data.
(c) Define any input data required in subroutine UREAL and transfer to UREAL by means of labelled common block.
(d) Call SEEK 1 to initiate search.
(e) Call ANSWER to give printed output.
(f) RETURN, END.

Method of calculation

The flowchart is given in figure 7.1. The Hooke and Jeeves search procedure follows as outlined in 3.3.3. The search is carried out until a maximum number of cycles have been exceeded or until no improvement can be found with all the step lengths less than their user specified minimum. It is possible for this type of method to hang up on constraints or to achieve a local rather than global optimum. For this reason the final optimum is checked by generating random values in the vicinity of the assumed optimum. If an improved point is found it becomes the new base point for another search procedure.

Optimization of a heat exchanger train

A heat exchanger sequence is to be designed comprising a boiler, 'stage 1', a boiler feed water preheater, 'stage 2', and a cooler, 'stage 3'. A gas stream is to be cooled from 800 to 35°C, for a flow, F, of 10 000 kg/h and 8 400 h/annum. Defining each stage as i, heat transfer equations and data used are as follows.

Design equations

The heat transfer at stage i is given by:

(1) $Q_i = U_i A_i (\Delta T_i)_{lm}$—overall heat balance

where
$$(\Delta T_i)_{lm} = \frac{(T_{i-1} - t_{io}) - (T_i - t_{ii})}{lm\left[\dfrac{T_{i-1} - t_{io}}{T_i - t_{ii}}\right]}$$

(2) $Q_i = FC_{hi}(T_{i-1} - T_i)$—balance over hot stream.

(3) (a) $Q_1 = \lambda W_1$—for waste heat boiler.

(b) $Q_i = W_i C_{ci}(t_{io} - t_{ii})$—for stages 2 and 3,

where the nomenclature is defined in figure 7.2.

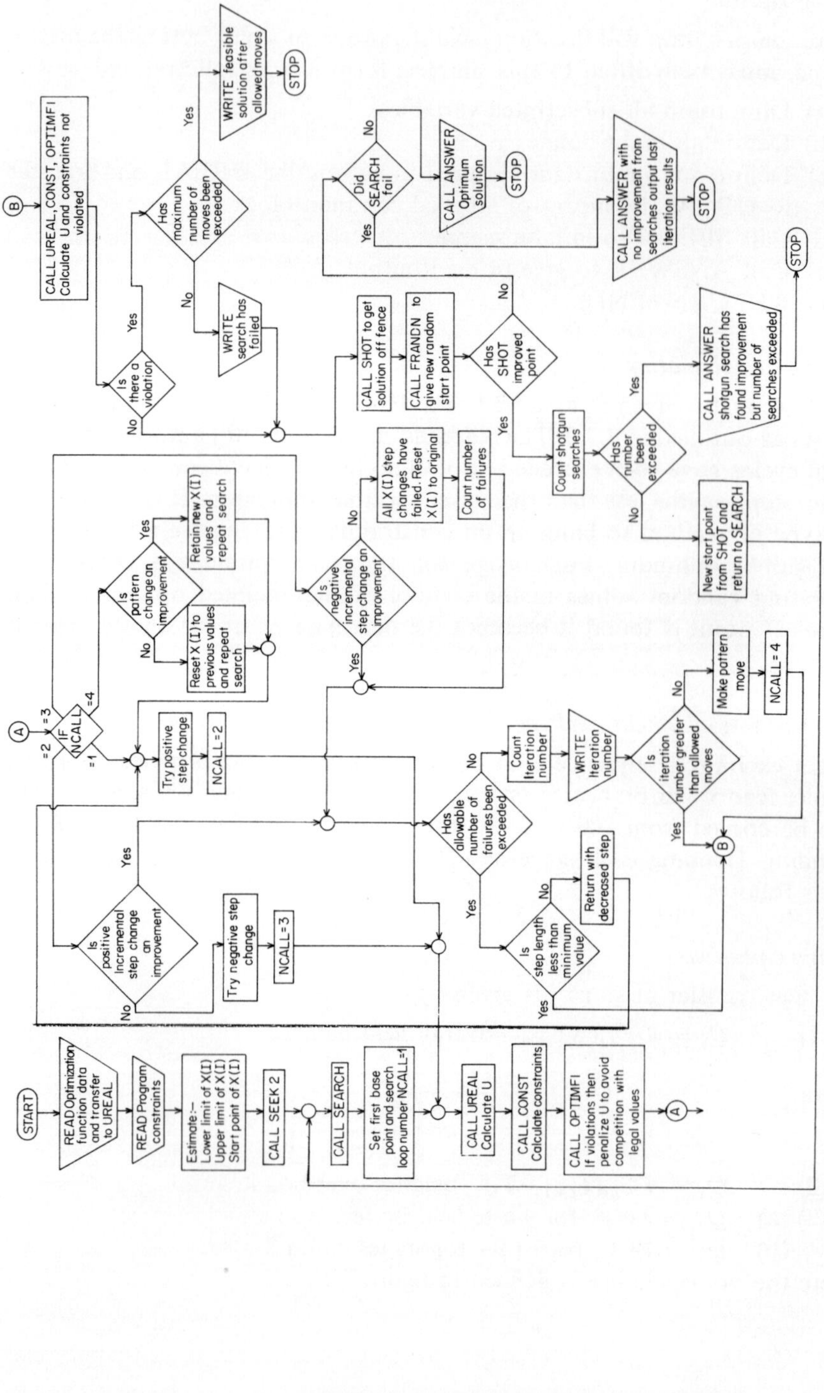

Figure 7.1 Flowchart of Hooke and Jeeves search.

Cost information is available as follows.
The capital cost of the heat exchangers is correlated against surface area as follows:

Boiler $I_E = 480$ A1 (exp 0.48)
B.F.W. $I_E = 400$ A2 (exp 0.48)
Cooler $I_E = 436$ A3 (exp 0.48)

where I_E is the delivered cost (£),
A is the surface area (m).

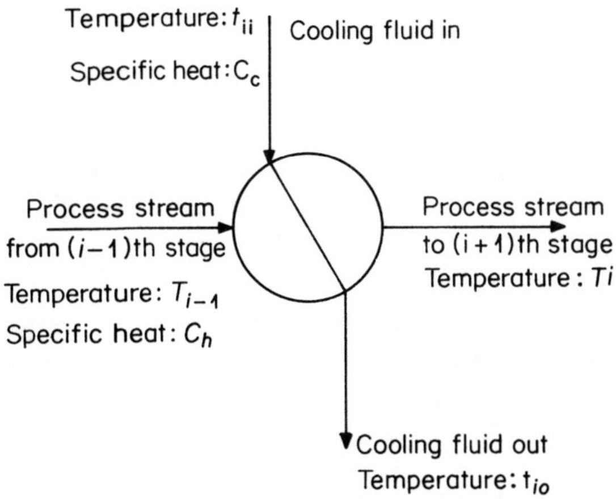

Figure 7.2 Nomenclature for exchanger sizing, ith exchanger.

Happel's formulation (1958) is used to derive installed costs, i.e.

$$\Sigma I_F = (I_E + \Sigma F_I I_E)(1 + \Sigma F_D)$$

where ΣF_I are indirect cost factors,
ΣF_D are direct cost factors,
ΣI_F is the installed cost (£).

Only factors which change with size of exchangers need be considered for optimization purposes, for instance instrumentation may be neglected except for the trivial case of omitting one exchanger. In this example factors were assumed as

direct: labour 0.15, piping 0.15, foundations 0.1, insulation 0.1 (not cooler),
indirect: overheads 0.2, engineering 0.05, contingency 0.5.

The cost function, U, for the particular depreciation, and minimum acceptable return is equivalent to

$$U = 0.55\, I_F + C_V$$

The overall cost function is given by the sum of

$U_{\text{boiler}} = 0.060\ A1\ (\exp 0.48) - 0.0014 W_1$
$U_{\text{b.f.w}} = 0.052\ A2\ (\exp 0.48) - 0.0002 W_2$
$U_{\text{cooler}} = 0.052\ A3\ (\exp 0.48) + 3 \times 10^{-7}\ W_3$

Data relevant to the problem is given in table 7.2. The unknowns are the process stream temperatures T_1 and T_2, and if these values are defined the various flow rates will be obtained.

Symbol		Stage 1	Stage 2	Stage 3
t_{ii}	Cooling fluid in (C)	240	100	20
t_{io}	Cooling fluid out (C)	240	240	35
U	Overall heat transfer coefficient (kcal $m^{-2}h^{-1}C^{-1}$)	300	300	400
C_c	Specific heat of cooling fluid (kcal $kg^{-1}C^{-1}$)	1.0	1.0	1.0
C_h	Specific heat of hot stream (kcal $kg^{-1}C^{-1}$)	0.60	0.57	0.55
J	Latent heat of cooling fluid (kcal kg^{-1})	670	—	—

Table 7.2 Data for optimization of heat exchangers.

Details of the programs constructed are given in program 7.2. The MASTER segment incorporated the estimated upper bound on $X(I)$ as 800°C for $I = 1, 2$. The estimated lower bound on $X(I)$ is 241°C for XMIN(1) and 101°C for XMIN(2). Feasible starting values of $X(I)$ are taken at the midpoint of the range.

The subroutine UREAL contains the heat balances and permits the objective function to be evaluated.

Subroutine CONST is written where the solution is constrained as follows:

$X(1) \geqslant 241\ (t_{1i})$ $\quad X(2) \geqslant 100\ (t_{2i})$
$X(1) \leqslant 800\ (T_0)$ $\quad X(2) \leqslant 800\ (T_0)$

N—no. of variable = 2
NCONS—no. of inequality constraints = 4

and

PHI(1) = X(1)−100.
PHI(2) = 400.−X(1)
PHI(3) = X(2)−200.
PHI(4) = 600.−X(2)
PHI(5) = X(2)−101.

The job was processed on an I.C.L. 1904 and the mill time was 10 seconds for the convergence to a solution. Approximately 250 alternatives were examined automatically. The recommended temperatures were

$K(1) = T(1) = 277.75°C$
$T(2) = 113.17°C$

```
      SUBROUTINE UREAL (X,U)
      DIMENSION X(2)
      COMMON/HEAT/U1,U2,U3,RATE,EX1IN,EX2IN,EX3IN,EX1OUT,
     1 EX2OUT,EX3OUT,SH1,SH2,SH3,LAT,TEMP1,TEMP4
      TEMP2=X(1)
      TEMP3=X(2)
      TMEAN1=((TEMP1-EX1OUT)—(TEMP2-EX1IN))/ALOG(ABS
     1 ((TEMP1-EX1OUT)/(TEMP2-EX1IN)))
      TMEAN2=((TEMP2-EX2OUT)-(TEMP3-EX2IN))/ALOG(ABS
     1 ((TEMP2-EX2OUT)/(TEMP3-EX2IN)))
      TMEAN3=((TEMP3-EX3OUT)-(TEMP4-EX3IN))/ALOG(ABS
     1 ((TEMP3-EX3OUT)/(TEMP4-EX3IN)))
      Q1=RATE*SH1*(TEMP1-TEMP2)
      Q2=RATE*SH1*(TEMP2-TEMP3)
      Q3=RATE*SH3*(TEMP3-TEMP4)
      AREA1=Q1/(U1*(ABS(TMEAN1)))
      AREA2=Q2/(U2*(ABS(TMEAN2)))
      AREA3=Q3/(U3*(ABS(TMEAN3)))
      FLOW1=Q1/LAT
      FLOW2=Q2/(EX2OUT-EX2IN)
      FLOW3=Q3/(EX3OUT-EX3IN)
      COST1=(0.06*(AREA1**0.48))-(0.0014*FLOW1)
      COST2=(0.052*(AREA2**0.48))-(0.0002*FLOW2)
      COST3=(0.052*(AREA3**0.48))+(0.000003*FLOW3)
      U=COST1+COST2+COST3
      WRITE(6,60)TEMP2,TEMP3,U
   60 FORMAT(' TEMP2=',E16.8,' TEMP3=',E16.8,' U=',E16.8)
      RETURN
      END
      SUBROUTINE ANSWER(U,X,PHI,PSI,N,NCONS,NEQUS)
      DIMENSION X(2),PHI(5),PSI(1)
      COMMON KO, NNDEX
C     THIS SUBROUTINE IS USED MERELY TO OUTPUT FINAL
C     SOLUTION IN A STANDARD FORM. IF AN OPTIMUM IS
C     NOT REACHED (KO=1) THEN THE RESULTS AT THE LAST
C     ITERATION MAY BE PRINTED OUT.
      CALL UREAL(X,U)
      IF(KO.EQ.0)GOTO1
      WRITE(6,18)
      WRITE(6,19)U
      GOTO2
    1 WRITE(6,20)
      WRITE(6,21)U
    2 WRITE(6,22)(I,X(I),I=1,N)
      IF(NCONS.EQ.0)GOTO3
      CALL CONST(X,NCONS,PHI)
      WRITE(6,23)
      WRITE(6,24)(I,PHI(I),I=1,NCONS)
    3 IF(NEQUS.EQ.0) GOTO 30
      CALL EQUAL(X,PSI, NEQUS)
      WRITE(6,25)
      WRITE(6,26)(I,PSI(I),I=1,NEQUS)
```

Program 7.2 Optimization of a heat exchanger train.

```
18 FORMAT(1H,16X,25HRESULTS AT LAST ITERATION,/)
19 FORMAT(29X,3HU=,E16.8//)
20 FORMAT(1H1,21X,22HOPTIMUM SOLUTION FOUND,/)
21 FORMAT(20X,12HMINIMUN U =,E16.8//)
22 FORMAT(25X,2HX(I2,3H) =,E16.8)
23 FORMAT(1H,22HINEQUALITY CONSTRAINTS)
24 FORMAT(23X,4HPHI(I2,3H) =,E16.8)
25 FORMAT(1H,22H EQUALITY CONSTRAINTS)
26 FORMAT(23X,4HPSI(I2,3H) =,E16.8)
30 RETURN
   END
   SUBROUTINE CONST(X,NCONS,PHI)
   DIMENSION X(2),PHI(5)
   PHI(1)=X(1)-X(2)
   PHI(2)=800.-X(1)
   PHI(3)=X(1)-241.
   PHI(4)=800.-X(2)
   PHI(5)=X(2)-101.
   RETURN
   END
```

Program 7.2 Optimization of a heat exchanger train—*continued.*

This is, of course, to a greater accuracy than warranted for this problem and represents inefficient programming on our part in defining the required accuracy.

As a further exercise a separate FOCAL program was written for a PDP 8 to permit evaluation of the objective function. The search routine was not added. Students were invited to find the optimum after being given an outline of the Hooke and Jeeves search principles. On average 50 alternatives were examined to achieve the same accuracy of solution to the nearest °C. Because of thinking time the problem took some 20 minutes to resolve.

This is an extremely simple example of the use of this program and the reader may consider the time required to solve problems containing three values of (I)I without the benefit of the program for automating the search procedure.

7.4 Large systems

Larger systems may be tackled in the same manner. The formulation of the problem is expressed in section 3.3 as

$$\max_{q}\{\phi(\boldsymbol{p};\boldsymbol{q})\,|\,g(\boldsymbol{p}:\boldsymbol{q})\geqslant 0\} = \phi^*(\boldsymbol{p}) = \phi(\boldsymbol{p},\boldsymbol{q}^*)$$

where ϕ is the objective function whose value is determined by the decision variable vector $\boldsymbol{q}$, and the parameter vector $\boldsymbol{p}$.

Hughes (1969) has described the operation of CHEOPS which contains a number of algorithms which solve any problem formulated in this fashion. Table 7.3 shows the usable problem size for the optimization algorithms used. The objective function is frequently non-linear and the references in the table provide an introduction to this problem. Box, Davies and Swann (1969) describe numerous non-linear optimization techniques and Sargent (1967, 1968, 1972) has regularly reviewed the literature as well as indicating several of his procedures.

From the literature these algorithms would appear to have been more used in the optimization of existing plant than for individual plants. Friedman and Pinder (1972) have for instance modelled a gasoline polymerization plant. They are also important in the development of routines for category C programs and repetitive small systems, i.e. heat exchanger and column trains. When simultaneous solution of a large projected system is being attempted, the algorithms are more readily applied.

	LP	MAP	GP	PPNL	DA
Objective function	L	NL	NL	NL	NL
Decision variables					
linear	10^4	500	100	600	40
non-linear				60	
Bounds			200	120	80
Constraints } Equations	2 000	1 000	200	2 000	20

where L is linear
NL is non-linear
LP —linear programming
MAP —Griffith and Stewart (1961)
GP —Gradient projection
PPNL—Ornea and Eldredge (1965)
DA —Direct ascent

Table 7.3 Problem size for optimization algorithms.

Otherwise to evaluate the best process route for a proposed plant, overall algorithms may not be required. Instead, runs through the system can be carried out varying parameter as the user so desires. The problem in such use is not only one of computation cost but also of formulation time and data organizing. It is necessary to keep the model as simple as is consistent with the problem. Thus if a heat exchanger is not a key item (it rarely is), use overall heat transfer correlations at this optimization stage. Remember cost information is rarely accurate. Do not specify all components. As in distillation some components are the 'key' to the problem. Try to use a good base case and study marginal changes preferably isolating these to key areas. Unit subroutines which involve many iterations for their solution are undesirable within the mass balancing routines and simultaneous solution is desirable. In some cases one should not automatically go to a packaged process simulator to evaluate alternatives but spend time in formulating the mass balance routine as a one-off job. Remember that implicit constraints may arise which are limitations on values of items in the process flow but are not optimization parameters, i.e. liquid below its boiling point.

The problem is usually studied as deterministic but this is itself an approximation. This should be remembered when tackling the problem and be reflected in the desired accuracy of solution. Key parameters should be varied to optimistic and pessimistic values before finalizing on the best solution.

7.5 Special procedures

There are a number of techniques which although providing a framework in which optimization might be applied do so in a manner which tends to avoid the use of the above algorithms, and yet reduce the number of cases attempted. These include dynamic and branch and bound integer programming techniques.

Dynamic programming attempts to optimize the plant in stages such that whatever the initial state and the initial decision are ,the remaining decisions constitute an optimal policy with regard to the state resulting from the first decision (Bellman, 1957). The calculation consequently is carried out in reverse manner to the flow of information.

Algebraically at the nth stage

$$\boldsymbol{p}_n = J_n(\boldsymbol{p}_{n+1};\boldsymbol{q}_n)$$

meaning that given $\boldsymbol{p}_{n+1}$ and $\boldsymbol{q}_n$ we can calculate $\boldsymbol{p}_n$. Consider the sequence:

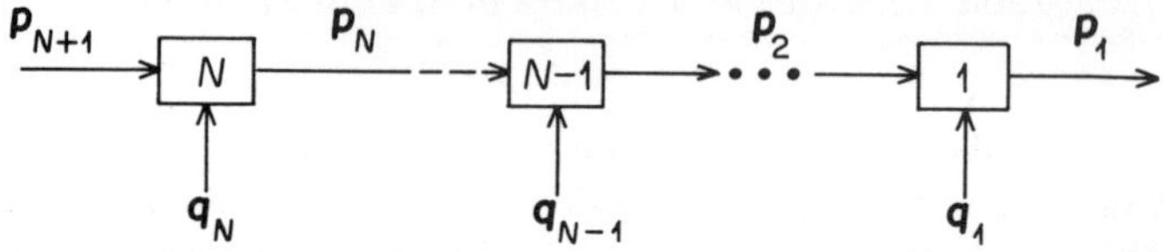

For the final stage if the feed is specified, $\boldsymbol{p}_2$, then the principle of optimality dictates that the decision vector $\boldsymbol{q}_1$ is selected so that the objective function, y, is an optimum for that feed. Denoting the optimum by M_1, then

$$M_1(\boldsymbol{p}_2) = \underset{\boldsymbol{q}_1}{\text{opt}} \{y_1(\boldsymbol{p}_2;\boldsymbol{q}_1)\}$$

and for the last two stages

$$M_2(\boldsymbol{p}_3) = \underset{\boldsymbol{q}_1\boldsymbol{q}_2}{\text{opt}} \{y_1(\boldsymbol{p}_2;\boldsymbol{q}_1)+y_2(\boldsymbol{p}_3;\boldsymbol{q}_2)\}$$

or as $\boldsymbol{q}_1$ is selected to optimize y_1 for a given $\boldsymbol{p}_2$

$$M_2(\boldsymbol{p}_3) = \underset{\boldsymbol{q}_2}{\text{opt}} \{M_1(\boldsymbol{p}_2)+y_2(\boldsymbol{p}_3;\boldsymbol{q}_2)\}$$

Thus in general

$$M_n(\boldsymbol{p}_{n+1}) = \underset{\boldsymbol{q}_n}{\text{opt}} \{M_{n+1}(\boldsymbol{p}_n)+y_n(\boldsymbol{p}_{n+1};\boldsymbol{q}_n)\}$$

where $p_n = \tau_n(p_{n+1}; q_n)$, $n = 1, 2, \ldots, N$.

This case is extremely simple because no branched streams enter the problem. Aris (1964) has considered some possible cases which provide an introduction to the greater problem. A simple example is when two branches occur:

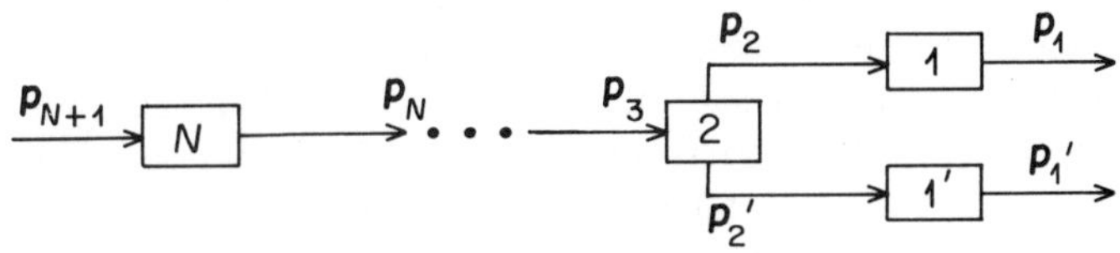

The maximum returns from the separate branches can be calculated by

$$M_2(p_3) = \max_{q_2} M_1 [M_1(p_2)+M_1'(p_2')+y_2(p_3; q_2)]$$

and

$$M_n(p_{n+1}) = \operatorname*{opt}_{q_n} [M_{n-1}(p_n)+y_n(p_{n+1}; q_n)],$$

$n = 3, \ldots N$.

It is self-evident that using this strategy it is possible to reduce the number of calculations as compared with direct search, table 7.4. Branching makes automating the procedure difficult and the storage requirement is considerable for keeping all the cost data derived from calculating the various costs associated with different actions in processing specific input streams. Consequently the strategy is only useful for a few cases in real problems. Rudd and Watson (1968) have simplified the procedure by calculating the data at each stage and assuming that a continuous cost curve may be obtained from these results. The calculation may then be carried out manually selecting the optimum off the curve at each stage. Their book is recommended as an introduction to this strategy.

An alternative to dynamic programming is heuristic programming in which having arrived at a decision point, the decision is made based on rules

		Number of components		
		2	5	10
Number of state variables between two components	1	0.5	13	22 000
	3	5×10^{-4}	0.15	220
	5	5×10^{-5}	1.5×10^{-3}	2

Table 7.4 Comparison of dynamic programming and direct search (Rudd and Watson, 1968). Ratio of work by direct search to dynamic programming, assuming one design variable per stage.

of thumb which circumvent analysis of the problem. Improper optimization may be made but this is unavoidable if the number of alternatives as in games such as chess are so great. The program is built up by discovering the path which for most cases leads to the best value attainable given various conditions input at the decision event.

Such a system is analagous to that employed by the process designer when he synthesizes the process by using his experience to select which unit operation is most appropriate for a given separation. Programming is simple, it is the formulation of the structuring criteria which is difficult. For examples of the technique see Masso and Rudd (1969), for work relating to synthesis of a process and who have produced networks permitting the selection of the correct heat exchanger type for a given task.

Very similar to the above approach is the branch and bound method in which an attempt is made to branch from a difficult problem to a simpler

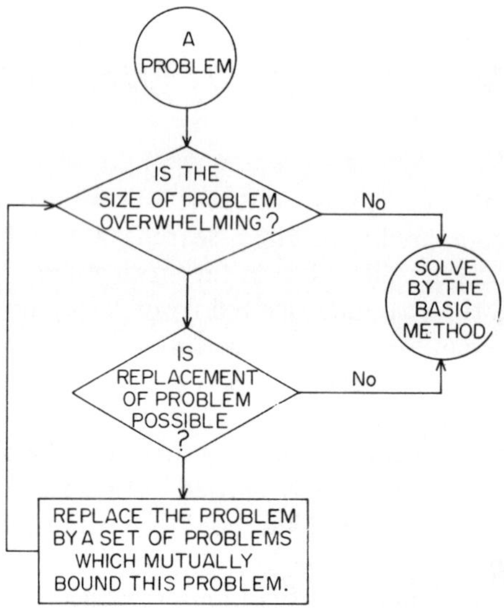

Figure 7.3 Algorithm for branching strategy (Lee, Masso and Rudd, 1970).

problem which bounds the original, figure 7.3 (Lee, Masso and Rudd, 1970). The criteria for success is that if the optimal solution to the original problem were available and applied to solve the new alternative route then the problem should not infringe any technical constraints along this route. The new problem then bounds the original problem. An example of this approach is to relax a constraint on the original problem and then solve the new problem. It may be that the original constraint is violated but the experience gained may permit of a new effort to branch successfully. The strategy is used for instance in the design of distillation columns using the Fenske–

Underwood technique. Instead of separating one compound from, say, six compounds the problem is treated as separating two key components. As for dynamic programming, application can result in excessive computation and storage problems when automating the strategy.

An introduction to the use of such programs is provided by the publications of the larger computer manufacturers. Thus International Computers Ltd (I.C.L.) regularly publish bulletins on their mathematical programming technical software, and, for instance, their linear programming mark 3 manual contains a useful introduction to integer programming.

Notation	Definition	Section
A	area of surface	general
c	cost of steam/unit of heat	7.1
C	specific heat	7.3
C_r	variable costs	7.3
$\mathbf{d}$	depreciation	7.1
F_I	indirect cost factor	7.3
F_D	direct cost factor	7.3
F	hot fluid flowrate	7.3
g	investment grant	7.1
$\mathbf{g}$	constraint limitations vector	7.4
H	upper optimum limit	7.3
i_m	minimum acceptable return	7.1
I_E	delivered costs	7.3
I_F	installed costs	7.3
I	investment	7.1
k	thermal conductivity	7.1
L	lower optimum limit	7.3
M	optimum of objective function	7.5
N	number of variables	7.3
$\mathbf{p}$	parameter vector	7.4, 7.5
$\mathbf{q}$	decision variable vector	7,4, 7.5
Q	heat duty	7.3
T, t	temperature	general
t	taxation	7.1
U	economic objective function	7.1
U	overall heat transfer coefficient	7.3
W	cold fluid flowrate	7.3
x	insulation thickness	7.1
x	mole fraction liquid	7.2
x	variable	7.3

Greek symbols		
Δ	difference	general
α	relative volatility	7.2
Θ	Underwood parameter	7.2
ϕ_K	inequality constraints function	7.3
ϕ	objective function	7.4
λ	latent heat of vaporization	7.3
τ	function	7.5

Subscripts

C	cold fluid	7.3
C_i	cold fluid in	7.3
F	feed	7.2
h_i	hot fluid in	7.3
h	hot fluid	7.3
i	heat exchanger stream stage	7.3
i	component	7.2
ii	ith stage in	7.3
io	ith stage out	7.3
lm	logarithm mean	7.3
N	variable identifier	7.3
N, n	stage number	7.5

8 General Use of Computers in the Chemical Industry

8.1 General philosophy

Having examined the main types of program being used by process engineers, let us consider the general use of computers in the chemical industry concentrating on activities impinging on process engineering.

A few examples involving the use of the computer in relation to chemical engineering design have been given earlier. What are the advantages to the engineer in such use? We have seen that the computer is a large calculating machine capable of extremely rapid calculations. On the other hand, the logical decision processes are programmed into the machine. Although in theory it is possible to build every program available into a vast design simulator, such a method would involve an intolerable computing time, even if a machine of sufficient storage capacity existed. Also, as we are evaluating a specific system it may well be that in defining the elements of this system a sufficiently fundamental level has not been used.

As the engineer selects where to use the machine, he should first consider the advantages of such use. Clearly the machine can relieve the user of the chore of calcualtions; this enables the user to increase his productivity, carry out work which he finds more interesting and avoid calculation errors (assuming the programs have been sufficiently proven).

The computer can carry out calculations which hitherto have been too tedious to solve by hand methods. This results in a greater accuracy by eliminating approximate methods at the choice of the user.

A greater number of cases may be considered, such as adjusting the parameters of an objective function, or considering a greater number of alternative items of equipment if these can be incorporated in a logical plan.

The large filing capacity of the computer can be used as a source of data, such as physical properties of materials, notes on previous designs, cost information, a list of equipment to order.

The computer may control a repetitive series of operations without tiring and to a greater accuracy due to its speed of thought; control of plant is an obvious example. Robbins (1970) has discussed these aspects in greater detail.

8.2 Category of use

The process engineer has many programs available to help him in his work and they may be categorized as in table 8.1.

The task of producing and operating a chemical plant involves several stages and each contains several largely independent areas of endeavour. The stages have been detailed by Wells (1973) and a study of this book permits further judgement as to the type of program to use in the specific circumstances which are dictated by the appropriate, economic objective function.

In this chapter we will concentrate on discussing the applications of computer aids in the various areas of endeavour.

8.3 Development

During the development stages of a chemical plant it is necessary to make assessments of the costs of the full scale plant. The programs, if written in general form, permit many alternative designs to be investigated quickly. Using such programs can also indicate areas where the process parameters are ill defined, thereby permitting experimental work to be initiated with top priority.

Computer generation of new processes is very much in its infancy and it is hoped that developments will achieve success. However, the vast range of actions at any decision event is so great as to make this unlikely except for experimental purposes.

8.4 Process design

We have earlier discussed process design with respect to process simulators and unit computations. It is emphasized that normally the operating company will utilize the services of contractors for the design and construction of plant and manufacturers of equipment. Thus the concept of a complete simulator which from a specific plant structure can define the exact specification of every pipe and item to be used is probably not economically viable. The specialist heat exchanger designer can by virtue of his improved programs probably design to closer tolerance of capacity; the fabricator can supply a unit from his standard range at lower price than for a custom-built item. Thus as well as considering the economic objective function of the process engineering exercise which he is carrying out, the engineer should also appreciate what work other specialists will carry out after he has finished his design.

At present some manufacturers of items such as distillation columns and heat exchangers are making use of well tried computer design methods

Category	*Purpose*	*Typical use*
Data programs	Analysis of experimental work. Provision of data for process design.	Equation and curve fitting. Physical property estimation. Calculation of phase equilibria.
Balance programs	Provide data for unit computations.	Heat and mass balancing.
Unit computations	Design of plant items.	Specification of columns, exchanger, reactors. Design of specialized units.
Unit process models	Design and optimization of process units.	Process simulation on simulator. Design of process package for integration in main process, i.e. purification units.
Plant models	Optimization of projected or existing process. Process simulation of existing unit.	Evaluation of return on capital of proposed units. Model effect of various operating changes to optimize process. Process plant profit and loss account.
Works and company models	Production planning and economic optimization.	Linear programming models.

Table 8.1 Classification of process programs.

either as developed by themselves or purchased from specialist designers in this field. The use of such designs obtained if one is a member of a particular 'club' seems a logical method of avoiding the expense of developing programs.

The development of improved models of units will continue. As far as the computational aspects are concerned there is some need for generalized solutions where the chemical engineering content and the numerical methods are kept in separate compartments. If the chemical engineering content is expressed in the form of a set of linear/non-linear equations then the appropriate method of solution might be called up as a subroutine. This is of course an oversimplification as it is necessary to know which numerical methods might succeed for a given formulation of the problem, and the simpler methods may readily be programmed. But it is not practicable for the general user to do other than call up the more complex numerical methods. On the latter, work is required to make available the best methods both with respect to ability to solve equations and computational efficiency.

The use of the computer as a file for process data becomes invaluable and is a ready example of relieving the process engineer of a most wearisome chore.

8.5 Mechanical design

Many designs stop short at process design, partly because of the nature of the industry. However, the process information is rarely taken by a manufacturer and input to a mechanical simulator to give a complete specification for manufacture. Instead, checks are made as to how the unit fits into the existing range of designs and hand/machine calculations used in conjunction with codes of practice to check on special details. However, note that the techniques are available to build up suitable programs and presumably these will in time be used. It should be noted that a major problem arises in obtaining accurate data as the labour costs of individual manufacturers vary and such costs change as a function of time. This can mean an excessive input of fresh data for a truly optimized design.

The use of modular construction can help manufacturers improve their production methods and this in turn leads to increased use of the computer for stock control, planning production, sales control. A side effect is improved reliability of parts and when failure does occur, availability of spares.

If these factors are integrated the case for a computer terminal becomes improved and mechanical simulation should increase in importance. After all, we are considering a systems hierarchy and if the process units are the elements of the process, the component parts are the elements of the process units. Is what is good for the goose, good for the gander?

Automated draughting procedures are feasible either for standard designs using digital plotters or for special cases with the aid of computer graphics.

The latter is only economic when there is a high level of design content associated with the drawing rather than as simply replacing a draughtsman.

The computer permits ready access to a file of previous designs and lists of manufacturers. This should result in greater efficiency of the project engineer. Once bid tabulations are received for equipment these should be analysed to balance operating and capital costs and this includes reliability. The build up of satisfactory data followed by mathematical analysis of the quotations is feasible by computer and should be incorporated.

For capital cost purposes the computer is of value as a file. Factorial methods for capital cost estimation are of course readily computerized. However, because of the problems associated with achieving accurate preliminary estimation before full site and piping studies are computed, and variance in manufacturer's profit and overheads, it is unlikely that fully automated computer methods will eliminate conventional methods linked with the computer as a file. If, as the cost of proposed plant is built up the actual costs are substituted, then the variance between estimate and reality emerges. This can lead to improved forecasting but understandably the holding of post mortems is not universally popular. A check on the forecast cost of plant items as used in the optimization exercise is desirable and the revised information updates the file. At the control stage of cost estimation the computer can give assistance in controlling costs.

8.6 Integrated specification of units

The greatest implication with respect to the use of computers to help control the project more efficiently arises with respect to the communication between design sections. Clearly this could result in the consequences of design change or cost change from the original estimate (as used for optimization) being rapidly highlighted. Aspects such as modifying process design to permit the use of cheaper or available materials could quickly be incorporated. Thus using the computer the integration of process and mechanical design is feasible. *Such a change would radically alter the operation of a design office and consequently it is appreciated that considerable doubt arises about its benefit.* An analogy between the use of rotary or piston engines in cars as from, say, 1960 probably parallels its development. Note too that when this change comes about process simulators or unit computations will have to be extended such that one of their outputs is complete specification of equipment to manufacturers.

Such integrated simulation should stop at the specification of the unit. Two breaks then occur; the first when the specification sheet is sent to manufacturer for tender, the second at the layout stage. A simplified project network is given in figure 8.1 which illustrates the main flows for the main design activities. These are classified as:

(a) Process flow
(b) Process specification
(c) Engineering flow
(d) Mechanical specification
(e) Detailed draughting of equipment
(f) Layout
(g) Pipe routing
(h) Pipe detailing
(i) Fabrication of pipework
(j) Fabrication of equipment

omitting in particular civil, structures and instrumentation.

If this is analysed it may be seen that the sections may be integrated together into larger unit blocks as in figure 8.2. This diagram is hardly a surprise as it duplicates many aspects of project organization, thus showing how these are based on reducing information flow between sections. The sections may be considered separately for economic justification.

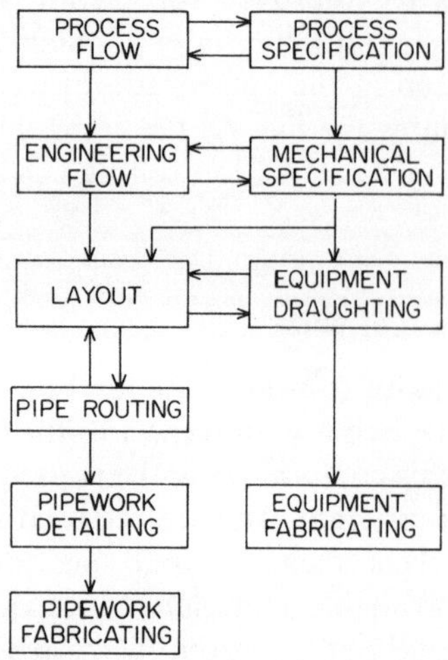

Figure 8.1 Simplified project structure.

Having carried out work on computer simulation of plant layout (Bush and Wells, 1972), along with other workers in this field (Leesley and Newell, 1972) we feel the construction of a plant model will remain a necessary stage in the design of a chemical plant. This prevents complete simulation of the overall system without access to a major activity outside the range of the machine. An additional break occurs when having specified equipment, outside manufacturers submit tenders for supply of units.

Thus for simulation of process and mechanical design it seems logical to extend the process simulation to incorporate engineering flow diagrams (E.F.D.) and mechanical specification stopping short at a complete specifica-

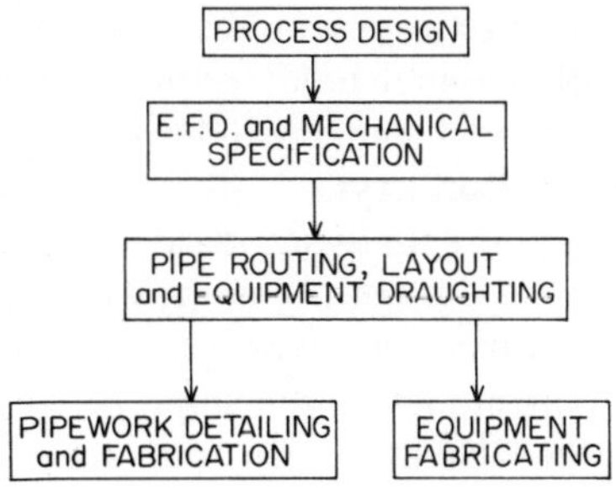

Figure 8.2 Design packages for the computer.

tion for manufacture with its attendant problems of detailed draughting. For layout an indication by the computer of a unit plot plant permits the plant modellers to make a sound start and improves estimation of piping costs.

Process simulation has earlier been discussed and after describing the computing hardware and software required we will discuss the other operations mentioned above.

1 *The computer configuration*

To carry out overall design simulation requires an extensive computing system usually consisting of host and satellites, figure 8.3.

Such a system might consist of a minicomputer and display unit linked

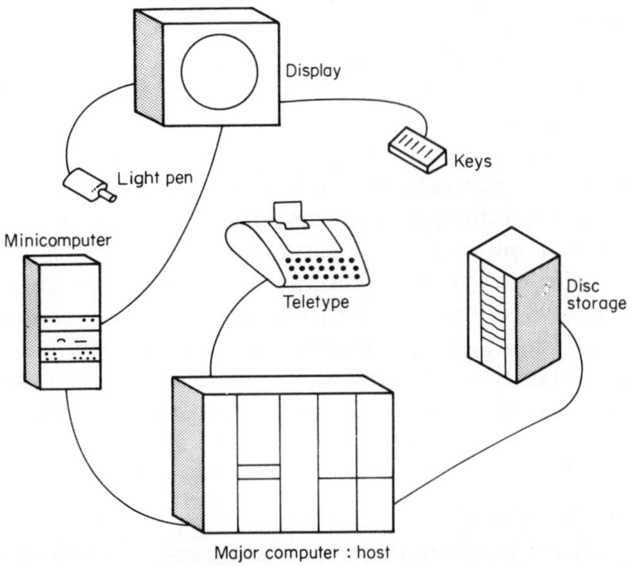

Figure 8.3 Computing system for overall design.

to a major computer. The program would be stored on disc and run in the larger processor, the minicomputer holding the interactive handling software which drives the visual display screen, the light pen, light buttons and tracking cross facilities and the function keys.

The use of function keys is kept as standard as possible between programs and consists mainly of basic draughting controls. The light buttons provide a key to the function and control of the program and therefore differ between programs or even different modes of the same programs. The light buttons appear as a menu on the display screen and upon selection provides the facilities outlined in the light buttons name. This may take the form of displaying a further menu of light buttons before the area of work to be tackled is entered.

The basic draughting controls provided as software to the display should permit the drawing of lines, circles and arcs of various types (i.e. chained or full) and their deletion. The lines should normally be automatically drawn vertical or horizontal except where a request is made to join two points. Auxillary aids permit production of captions to any appropriate size. Further light buttons permit the size of the picture to be readily changed by either magnifying or shrinking and the scale of the current view is displayed. Free manipulation of the picture is necessary. The drawing of exact distances from suitable input data and access to a library of standard shapes is essential for detailed and fast draughting purposes. Facility should be provided to make invisible unwanted lines at a particular stage of the work and different brightness of lines serves to avoid the necessity for various thickness of line as in a normal drawing. Copies may be taken either by photography or plotting in a subsidiary display file. The latter is slow but ensures that the whole of a drawing including portions not currently on the screen can be reproduced. Existing drawings can be stored in coded form in the backing store for ready access.

The benefits from such a computer system are easy modification of drawings and other data; fast access to calculations and applications programs; easy input of data and storage of previous designs; less chance of communication failure as most information is exchanged within the machine; automatic updating of schedules; giving process and project engineers automatic draughting facilities so he can readily complete the task without chance of errors in reproduction; greater capability of handling data.

The disadvantages include overstandardization; concentrating the design in the hands of the select few; and most importantly the high cost of time-sharing or hire of such interactive computing facilities.

2 *Engineering flow diagrams*

Engineering flow diagrams rather than the process flow diagram link the overall simulation of a chemical plant at the mechanical specification stage, figure 8.4. A graphical flow diagram program must permit the construction, modification, storage and retrieval of the diagrams which provide information for equipment design, layout, pipework estimating, capital cost estimating, control system design and utility requirements.

The main plant is usually sectioned together with diagrams of utilities. Continuations between line diagrams are shown by line reference and/or line content. For full details of the construction of these diagrams see Wells

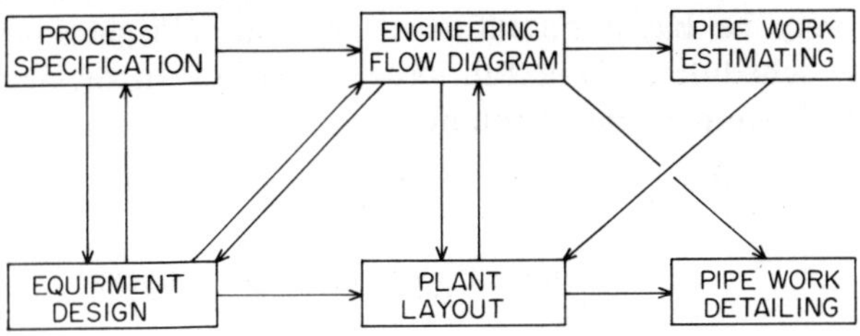

Figure 8.4 Integrated design flow diagram.

(1973). The diagram is constructed using the draughting facility of the computing system augmented by a library of symbols such as valves, pumps, vessels, etc. These may have indicated on them the normal attachment points, i.e. for a centrifugal pump as in figure 8.5. Connection between attachment points may be automated.

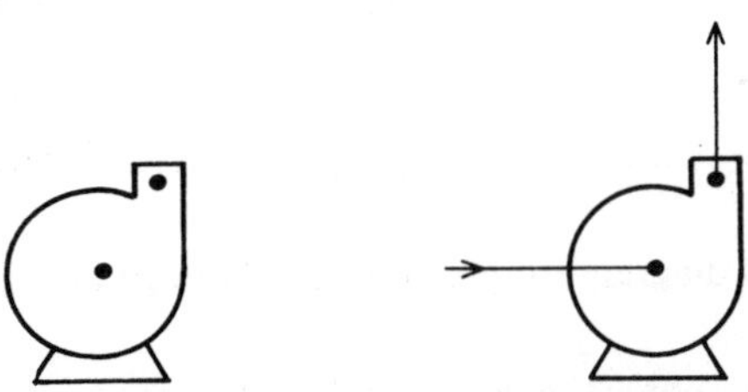

(a) (b)

Figure 8.5 Pump connections. (a) Before, (b) after connection.

Pipeline sizing should be carried out by this program. Whether or not exact floor locations are required depends on the automation of layout.

3 *Mechanical specification of equipment*

In general, considerable process design by simulators will be carried out without mechanical specification. However, once the design is close to being finalized it is possible to enter a phase for integrated equipment design. The principles are indicated for pressure vessels. The program is described in terms of the operating modes which are defined below and are integrated as in figure 8.6.

(a) *Initialization mode* The initial input to the program includes the facility to select from file equipment data and specification sheets and establish which subroutines will be used in the particular application.

(b) *Process specification mode* In the process design section information must be input to the program on design pressures and temperatures and process materials. The volume of the vessel may then be calculated. This information may be based on information from (1) sales and despatch, (2) analytical schedules, (3) a full process design incorporating process dynamics, or (4) simple rule of thumb.

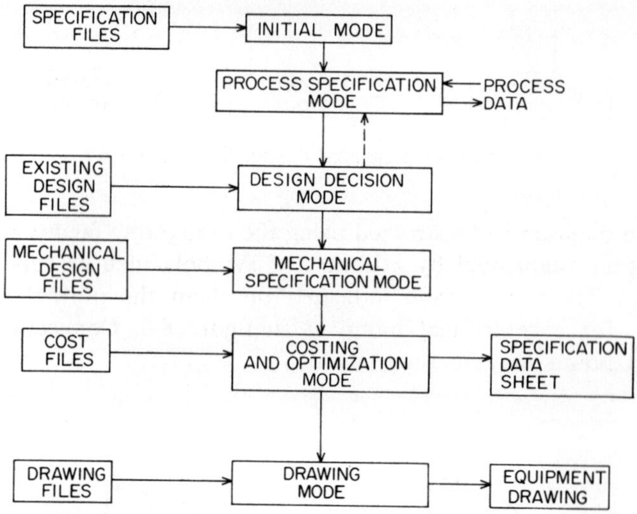

Figure 8.6 Structural diagram for pressure vessel design using computer graphics.

It is convenient to enter the design code at this stage. These might comprise B.S. 1500, B.S. 1515, A.S.M.E. and any special company codes. The material required for fabrication and the corrosion allowance should be entered, particularly if corrosive conditions exist.

(c) *Design decision mode* Based on this information the design decision mode is entered. What type of vessel is to be used, i.e. standard, spheres, floating roof, bullets, etc, for storage tanks? Is heuristic programming to be used for such selection? From the file of existing designs including external supply, can one be selected, and if so, is it to be modified or used in creating another? Any useful information on file is carried forward to the next mode.

(d) *Mechanical specification mode* We now enter the mechanical specification mode and in fact many designs could enter the program at this stage only, the earlier decisions being made using normal design procedures and the specification data sheet being filled in manually.

Diameter This may be selected on the basis of a length to diameter ratio which has in general proved to be in the economic range of $L : D$ ratio. It may be influenced by minimum instrumentation dimensions which set constraints or in exceptional cases by layout. Preferably, however, an optimization procedure should be written into the design. This may involve several iterations through the program before selecting the optimum design.

Shell thickness Calculated using stresses given in the appropriate codes. Stiffeners are incorporated as required and an iterative procedure is used. This will specify the material if not previously decreed. Otherwise, the data must be fed in manually from the codes. Thus for shells, the design stresses at design and room temperatures must be input.

Ends These are designed as per appropriate code. They include dished, hemispherical, bolted, welded and conical. Thus the appropriate type is selected by the user and the design information inserted as required.

Nozzles From the engineering flow diagrams the number of flanges is inserted either manually or by overall simulation. Each connection is then designed. Further allowance is made for any compensation and type of support. It is desirable to calculate the weight of the vessel. This information is printed together with a sketch of the vessel.

Access to these areas of work is made by selecting areas of work from light button menus which permit of highlighting such aspects.

(e) *Costing and optimization mode*

This stage of the program finally finishes with costing. This necessitates a detailed schedule of rates for building up the equipment cost. A typical build-up based on mythical values is given in table 8.2. Obviously this will apply to only very common materials and special corrections are required for different materials and upgrading cost information. It is rather impracticable for truly iterative calculations to be used for optimization and normally the user would insert alternative input information as he considers necessary.

This first cost is only approximate and developed for optimization purposes. Real costs are influenced by availability of dished ends, standard plate width, intermediate seams and welding quality. Overheads and profit are very dependent on business considerations of the vendor. Thus after obtaining a price from a manufacturer it is necessary to check with the estimates to ensure that an incorrect optimization was not made. Costing should not stop at equipment cost but by means of factors (Miller, 1963 and Wilson, 1971) should be further refined to installed equipment and combined in the process estimate costs. Care should be taken over the interpretation of the results and if necessary process input data should be modified. All optimization decisions should not violate Bellman's principle of optimality (1957).

Furman and Cheers (1971) have indicated a suitable algorithm for carry-

Reference: F-806, cyclohexane plant

Shell	Dia.	Thickness	Weight	Rate	Amount	Cost
Side	1 500	16	1 600	0.100	160	
End 1	1 500	30		Dished	100	
End 2	1 500	30		Dished	100	360
Nozzles						
1	600	10		150	50	
1	100	10		25	25	
3	80	10		20	60	
2	25	10		10	20	155
Blanks					35	
Instrumentation					0	
Manholes	1 500	16		104	104	
Saddles			300	0.200	60	
Lining m^2					0	
Painting 20.0 m^2				0.4	8	
Radiography 10 m				4	40	
Surface preparation 20.0 m^2				0.4	8	
Stress relief 3000 kg				0.01	30	
Total miscellaneous						285
Basic equipment cost						800
Overhead, drawing, delivery and profit at 20%						160
Total cost of vessel						960

Table 8.2 Cost analysis of pressure vessel.

ing out an exercise similar to that outlined above which could with ease be built into a computer graphics routine.

8.7 Project control

From specification of the unit the project enters the phase of civil, structural, piping design, etc. The automation of pipework estimating, detailing, costing and material control is described by Daniel (1971) for the ISOPEDAC system. Pipeline specifications and schedules of rates are processed to form project data banks. Specifications consist of sets of constant data for tubing, gaskets, butts, etc, which comply with pressure and temperature conditions of each pipeline. They also contain a schedule of rates for cost or manhours.

Pipework estimating is prepared after project sanction and involves making estimates from flowsheets and layout drawings of the footage of tube, number and type of items occurring in each line. The pipework design system produces for each batch and then for the whole project, total footage of tube and number of items, flanges, etc. The material totals for each batch are in a suitable form for ordering purposes. The procurement section returns to the computer the ordered quantities, supplies, order number costs and delivery dates. Detailed pipework data is subsequently prepared once pipework general arrangements or models are available.

A batch of pipework is prepared with job descriptions, specifications and schedule reference codes at the head. Each line of data contains the specification reference code, line reference origin as coordinates or description, directional abbreviations, dimensions and item codes for fittings. Output information consists of fabrication schedules with cutting and material lists; erection schedules, isometrics and material lists; bills of costs and manhours for fabrication, erection, lagging and paintwork; and a balance of orders against outstanding orders.

It is generally accepted that the advantages accrue mainly from the data collating capability of the machine. This greater efficiency in turn eases erection problems in reducing errors and improves planning of resources and sequence.

Overall project critical path planning is well established on the computer. McKenzie (1967) has described one system, AUTOPERT in detail. The system is important not only with respect to indicating delivery dates of equipment but also with respect to resource allocation, particularly for construction purposes. Figure 8.7 indicates the shapes of a project assuming various construction demands.

The advantage of the computer is that the system may rapidly be kept up

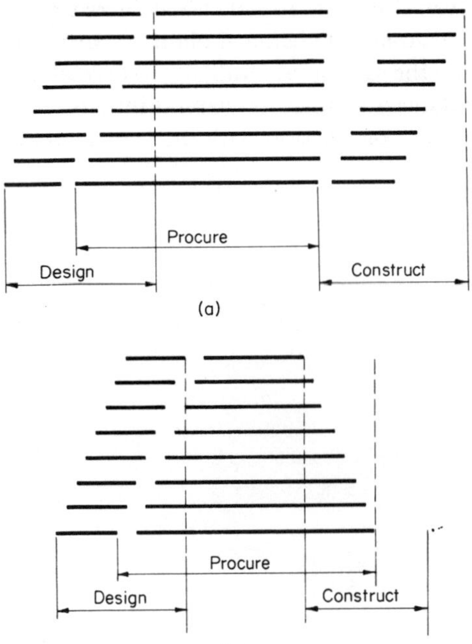

Figure 8.7 Resource distribution on a project. (a) Collapse of construction phase due to excessive build up of construction demands. (b) Well-balanced project (Gaussian distribution).

to date, the results printed in bar chart and histogram form. Resource allocation may be automated.

8.8 Operation

Production planning models using linear programming models are well documented and an example is given in Wells (1973). Data input to the computer is dependent on the particular programs available.

Commissioning of new plant should be planned by critical path scheduling, but day-to-day problems usually throw out the plan from an early stage.

Maintenance scheduling obviously can be planned. Within such schedules should be a data bank available for storing data on in-service reliability of equipment.

Technical studies will, of course, use self-standing programs for studies of equipment performance, debottlenecking, etc. Table 8.3 indicates possible sources of process improvement.

Economic simulation has also been discussed. Use of the modular system to give the plant, profit and loss account is recommended as being closer to local management's requirements.

Use of the computer for direct control of the plant is dependent on the economics of the given situation. It will clearly increase as the hardware costs are reduced. On an established chemical plant an early paper by Lawson and Youle (1966) suggests the savings are unlikely to exceed 5% of operating costs and this must be balanced against the capital investment. The latter may be obtained by analysis in conjunction with instrument manufacturers.

Main area	Subdivision	Questions and comments
Raw materials	Yield	Can we improve the yield by new catalyst, configurations, or operating conditions?
	Recycle	Does it overload any system? How could we eliminate build-up of contaminants?
Exit streams	Specifications on products	Can we change product specifications by changing concentration, phase, form? Are the specifications too rigid, realistic? What do the customers really want? What is the incremental cost of adding each specification?
	Effluents and vents	Is there waste thermal energy, mechanical energy, concentration? What is the amount and condition of effluents?
	Safety and control system	Details of flare, blowdown, and sampling.
	Start-up and shut-down	Volumes involved and disposal.
Time	Maintenance and downtime	When to replace? Maintenance costs, maintenance policy?
	Cycle time	What is it for reactors, filter, batch distillation?
	Problems causing downtime losses	What are the probable trouble areas? Anticipate corrective measures.
Utilities	Thermal energy	Is high-pressure steam used to produce energy and low-pressure steam for heating? Identify all sources of heat; how can they be utilized? Condensate return? Flash steam used?
	Electricity	How can the power factor be improved?
	Inert gas, water, steam, refrigeration, compressed air	Can the material or the energy be utilized?
Labour		Good relations, cooperation with plant tests, good housekeeping? Safety conscious?
Operating conditions		What is the effect of changes in temperature, pressure, concentration, flow rate, bleed-off rate, recycle rate, catalyst charge size, rpm?
Capacity	Increased	How many major bottlenecks exist?
	Reduced	Are there any operating problems when plant capacity is reduced?
Minor processing alternatives		New piping, packing, material of construction, equipment, additional coolers?

Table 8.3 Checklist for the identification of sources of process improvement based on Crowe and coauthors, 1971.

References

ALLEN, D. (1972). *Capital cost evaluation*. Inst. Chem. Eng., London.

AMUNDSON, N. R., PONTINEN, A. J. and TIERNEY, J. W. (1959). *J. Am. Inst. Chem. Eng.*, **5**, 295.

ANDREW, S. M. (1964). *Trans. Inst. Chem. Eng.*, **47**, T79.

ARIS, R. (1964). *Discrete dynamic programming*. Blaisdell, New York.

BARRON, D. W. (1963). *Computer J.*, **6**, 134.

BECKETT, K. A. (1969). *Chem. Engr.*, **47**, 163.

BELLMAN, R. (1957). *Dynamic programming*. Princetown University, Princeton, New Jersey.

BEVERIDGE, G. S. G. and SCHECHTER, S. (1970). *Optimization: theory and practice*. McGraw-Hill, New York.

BOX, M. J., SWANN, W. H. and DAVIES, D. (1969). *Non-linear optimization techniques*. Oliver and Boyd, Edinburgh.

BURGESS, G. and WELLS, G. L. (1973). 5th I.F.I.P. Conference on Optimization.

BUSH, M. J. and WELLS, G. L. (1971). *Brit. Chem. Eng.*, **16**, April, 325.

BUSH, M. J. and WELLS, G. L. (1972). Design decision and the computer, *Inst. Chem. Eng. Symp. Ser.*, No. 35, 2:15.

CAPITAL COST ESTIMATION (1969). Inst. Chem. Eng., London.

CHEERS, R. F. C. and FURMAN, T. T. (1971). *Brit. Chem. Eng.*, **16**, 478.

CHRISTENSON, J. H. (1970). *J. Am. Inst. Chem. Eng.*, **16**, No. 2, 177.

CROWE, C. M., HAMIELEC, A. E., HOFFMAN, T. W., JOHNSON, A. I., SHANNON, P. T. and WOODS, D. R. (1971). *Chemical plant simulation*. Prentice-Hall, New Jersey.

DANIELS, P. T. (1971). *Chem. Engr.*, August, 297.

DAVIDON, W. C. (1959). Variable metric method for minimization. A.E.C. Research and Development Report, ANL-5990 (Rev.).

EDIE, F. C. and WESTERBERG, A. W. (1971). *J. Chem. Eng.*, **12**, 114.

FENSKE, M. R. (1932). *Ind. Eng. Chem.*, **24**, 482.

FLETCHER, R. and POWELL, M. J. D. (1963). *Computer J.*, **6**, 163–168.

FLOWER, J. and WHITEHEAD, B. (1973). *Chem. Engr.*, April–May.

FRANKS, R. G. E. (1972). *Modelling and simulation in chemical engineering*. Wiley-Interscience, New York.

FRIDAY, J. R. and SMITH, B. D. (1964). *J. Am. Inst. Chem. Eng.*, **10**, 698.

FRIEDMAN, P. and PINDER, K. L. (1972). *Ind. Eng. Chem. Proc. Des. Dev.*, **11**, 4, 512.

GILLILAND, E. R. (1940). *Ind. Chem. Eng.*, **32**, 1220.

GRIFFITH, R. E. and STEWART, R. A. (1961). *Man. Sci.*, **7**, 379.

HAPPEL, J. (1958). *Chemical process economics*. Wiley, New York.

HENLEY, E. J. and ROSEN, E. M. (1969). *Material and energy balance computations*. Wiley, New York.

HIGMAN, B. (1967). *Comparative study of programming languages*. McDonald, London.

HIMMELBAU, D. M. and BISCHOFF, K. B. (1968). *Process analysis and simultation.* Wiley, New York.

HOLLAND, C. D. (1963). *Multicomponent distillation.* Prentice-Hall, Eaglewood Cliffs, New Jersey.

HOOKE, A. and JEEVES, T. A. (1966) in *Computers in engineering design,* vol. II. University of Michigan, Ann Arbor, Michigan.

HOOKE, A. and JEEVES, T. A. (1969). *Computer machines,* **8**, 312.

HOUTBY, D. K. and KHAN, A. U. (1972). Design decision and the computer, *Inst. Chem. Eng. Symp. Ser.*, No. 35, 3 : 3.

HUGHES, R. R. (1969). *Chem. Eng. Educ.*, Summer, 113.

HUTCHINSON, H. P. (1972). SYMBOL. University of Cambridge.

INTERNATIONAL BUSINESS MACHINES (1967). C.S.M.P., Form H20–036703, Prog. no. 360A–CX–16X.

INTERNATIONAL COMPUTERS LTD (1972). Linear programming, mark 3.

JOHNS, W. R. (1970). Use of Computers in Studies Preceding the Design of Plants, Symposium, E.F.C.E. Conference, April, Florence.

JOHNSON, A. I. and MORGAN, J. I. (1973). *Process Technology,* **18**, 25.

JOHNSON, A. I. (1970). GEMCS. University of Western Ontario, London, Canada.

KLUMPAR, I. V. (1970). *Chem. Eng.*, June 1970.

KLUMPAR, I. V. (1969). PROVES. Diamond Shamrock Corporation, Painesville, Ohio.

LAPIDUS, L. (1962). *Digital computation for chemical engineers.* McGraw-Hill, New York.

LAWSON, J. M. and YOULE, P. V. (1966). *Application of computers in chemical industry.* C.I.A., Brighton.

LEE, W., CHRISTENSON, J. H. and RUDD, D. F. (1966). *J. Am. Inst. Chem. Eng.*, **12**, 1104.

LEE, K. F., MASSO, A. H. and RUDD, D. F. (1970). *Ind. Eng. Chem. Fundamentals,* **9**, 48.

LEE, W. and RUDD, D. F. (1966). *J. Am. Inst. Chem. Eng.*, **12**, 1184.

LEESLEY, M. E. and NEWELL, R. G. (1972). Design decision and the computer, *Inst. Chem. Eng. Symp. Ser.*, No. 35, 2 : 15.

LEWIS, W. K. and MATHESON, G. L. (1932). *Ind. Eng. Chem.*, **24**, 494.

MARTIN, C. N. B. (1972). Design decision and the computer, *Inst. Chem. Eng. Symp. Ser.*, No. 35, 4 : 27.

MASSO, A. H. and RUDD, D. F. (1969). *J. Am. Inst. Chem. Eng.*, **15**, 10.

MCKENZIE, G. (1967). *Chem. Engr.*, 119.

MILLER, P. A. (1963). *Chem. Eng.*, Sept., 226.

MOTARD, R. (1971). CHESS. University of Houston.

MULLER, D. E. (1956). *Math. Table Aids Comp.*, **10**, 208.

NAGIEV, M. F. (1964). *The theory of recycle processes in chemical engineering.* Mcmillan, New York.

ORNEA, J. C. and ELDREDGE, C. G. (1965). *Inst. Chem. Eng. Symp.*, June, 4:15.

PETERSON, H. G. and SANSOM, F. J. MIMIC. S.E.S.C.A.

PONTON, J. (1972). DYNSYS. Edinburgh University, Edinburgh.

POWELL, M. J. D. (1964). *Computer J.*, **7**, 155–162.

RADIN, G. and ROGOWAY, H. P. (1965). NPL, **8**, 9.

ROBBINS, J. (1970). *Chem. Engr.*, October, CE298.

ROSEN, F. M. (1962). *Chem. Eng. Progress,* **58**, 10, 69.

ROSENBROCK, H. H. (1960). *Computer J.*, **3**, 175–184.

RUDD, D. F. and WATSON, C. C. (1968). *Strategy of process engineering.* Wiley, New York.

SARGENT, R. W. H. (1967). *Chem. Eng. Progress,* **63**, 71.

SARGENT, R. W. H. (1968). *Chem. Engr.*, **10**, 424.
SARGENT, R. W. H. (1972). *Chem. Engr.*, **262**, 226.
SARGENT, R. W. H. and WESTERBERG, A. W. (1964). *Trans. Inst. Chem. Eng.*, **42**, T190.
SHAROE, J. (1972). *Technology foundation course, unit 13.* Open University Press, Bletchley.
SIDDALL, J. N. (1970). OPTISEP. McMaster University, Hamilton, Ontario, Canada.
SOUDERS, M., JR. and BROWN, G. G. (1934). *Ind. Eng. Chem.*, **26**, 98.
STAINTHORPE, F. P. and WHITEHAUSE, P. A. (1967). *Inst. Chem. Eng. Symp. Ser.*, No. 23, 182–192.
STOECKER, W. F. (1971). *Design of thermal systems.* McGraw-Hill, New York.
TABOREK, J. and PALEN, J. W. (1972). Design decision and the computer, *Inst. Chem. Eng. Symp. Ser.*, No. 35, 2 : 39.
THIELE, E. W. and GEDDES, R. L. (1933). *Ind. Eng. Chem.*, **25**, 289.
UNDERWOOD, A. J. V. (1945). *J. Inst. Petrol.*, **31**, 111; (1933) **25**, 289.
UNDERWOOD, A. J. V. (1948). *Chem. Eng. Prog.*, **44**, 603.
WANG, J. C. and HENKE, G. E. (1966). *Hydrocarbon Processing*, **45**, No. 8, 155.
WEGSTEIN, J. H. (1958). *Communications Association for Computing Machinery*, **1**, 9.
WELLS, G. L. (1971). *Inst. Fuel*, **44**, 606.
WELLS, G. L. (1973). *Process engineering with economic objective.* Leonard Hill, London.
WILSON, G. T. (1971). *Brit. Chem. Eng.*, **16**, No. 10, 931.
WILDE, D. J. (1964). *Optimum seeking methods.* Prentice-Hall, Eaglewood Cliffs, New Jersey.

Appendix
MANUAL FOR PRIMER

1 BASIS

PRIMER (PRocess Iterative Mass-flow EvaluatoR) is a simulation program for material balance calculations designed to assist in initial process design of projected plant or the modelling of process plant flows and economic simulation of existing plant. The following input data is required:

Flow rates and compositions of known streams.
Stream numbers into and out of each unit.
Calculation order of individual units.
Unit parameters for proportioning flows.
Desired convergence unit and required tolerance on recycle flows.

PRIMER adjusts the input data to give mass and molar flow rates of unknown streams in tabular form (see case study). Costs may be added as required. The method of solution is iterative.

2 FLOW

Each stream represents a continuous flow of material through the plant. It is possible to use additional dummy streams to convey information about a flow to another unit. Units are moles/hour. Provision is made to transmit temperature and pressure of the stream.

3 SPECIFYING FLOWS

An input stream is one whose value is independent of other streams in the process. Each of these streams must be identified and read into the system.

Values of the recycle streams may be fed to the process in order to effect improved convergence. Such use is optional.

The user has to list the name, molecular weight and identifier of each component. Each stream sequence must list components in a fixed order indicating zeros if no value of a particular component exists in a stream.

4 CALCULATION ORDER

The process may consist of either acyclic operation or cyclic operation(s) in which the output stream from one or more units returns to its origin. The user must establish whether recycle occurs and indicate accordingly.

The user establishes the order in which the computer is to calculate the individual operations. These are listed in proper order. At the end of any loop either a convergence promoter or a simple test of convergence must be made. If any further loops occur the loop counter must be reset back to one by the module LOOP. In the event of difficulty in establishing this sequence the program ORDER (Johnson, 1970) should be used independent of the simulator.

5 SPECIFYING UNITS

For each unit the user has to specify operation parameters such as the unit number, the total input and output streams and the identifier of each stream, unit type, i.e. junction, separator, the quantitative parameters for effecting change of input flows to the desired output flow values, and the tolerance in the case of convergence promoters.

The following operations are indicated in the attached listing.

CSCADE	cascade, i.e. absorber
JUNC1	junction of streams
LOOP1	resets loop value at zero
REAC1	reactor
SEPAR1	separator based on split of feed
TEST	checks tolerance on recycle streams
WEGST	Wegstein convergence promoter

Additional units may be added in conventional manner. Ensure the appropriate calling modification to MODULE is made.

The operating equations and the unit array format is listed in the form of comment statements in each subroutine. It is an aim of the programmers that the working of each subroutine can be read directly from the program without other information.

6 DATA PREPARATION

The system as dimensioned can process approximately 100 streams containing 8 components and having 50 units.

The sequence for data input is as tabulated in sets 1–7. Cards are punched as designated by *I* for integer numbers and *X* for real numbers. Each list is started by data cards giving the name &ALIST and ends with a card &END as indicated in chapter 6. Details of the NAMELISTS are given in table A.1. For further information see the case study.

7 PRINTOUT

The following printout is produced automatically: loop number, convergence information at each loop, final printout of mass balance, changes if parameters altered on subsequent runs.

Options are provided for further printout by setting a value to the KPRINT terms as listed in DLOAD.

8 DATA ALTERATION FOR SEVERAL RUNS

The number of runs through the program is set by specifying NRUNS at the start of the data input.

Following a run data may be reset by calling DALTER and reading DLIST where MOD identifies whether modifications are to be made to streams or modules or both. NMUN represents the total number of modifications to a unit and NMFL is the corresponding value for streams, 100 flows and 100 unit modifications may be made.

An example of these changes is incorporated in the case study.

9 LISTING AND EXECUTIVE

The EXECUTIVE routine of the program controls the calling of the data reading routine, DLOAD, the report writing routine, WRITE, the data modification routine, DALTER, and the units in the correct order of LIST (via MODULE), as shown in the flowchart, figure A.1. A listing is given in table A.2.

Following a run data may be reset by calling DALTER and reading DLIST where MOD identifies whether modifications are made to streams or modules or both.

10 CASE STUDY

In order to indicate the input format to PRIMER the example in figure A.1 is indicated in table A.3 for a five-component system. Assumptions can be read directly from this table. The system converged after loops to a tolerance on stream 10 of 0.001 for each component. The principal output is given in table A.4.

The NAMELIST sets are

Set 1	ALIST	Program description
Set 2	BLIST	Process description
Set 3	FLOWLIST	Initial stream information
Set 4	UNITLIST	Unit description
Set 5	DLIST	Number of alterations to initial data
Set 6	MFLLIST	Stream alterations
Set 7	MUNLIST	Unit alterations

Table A.1 Data preparation for PRIMER.

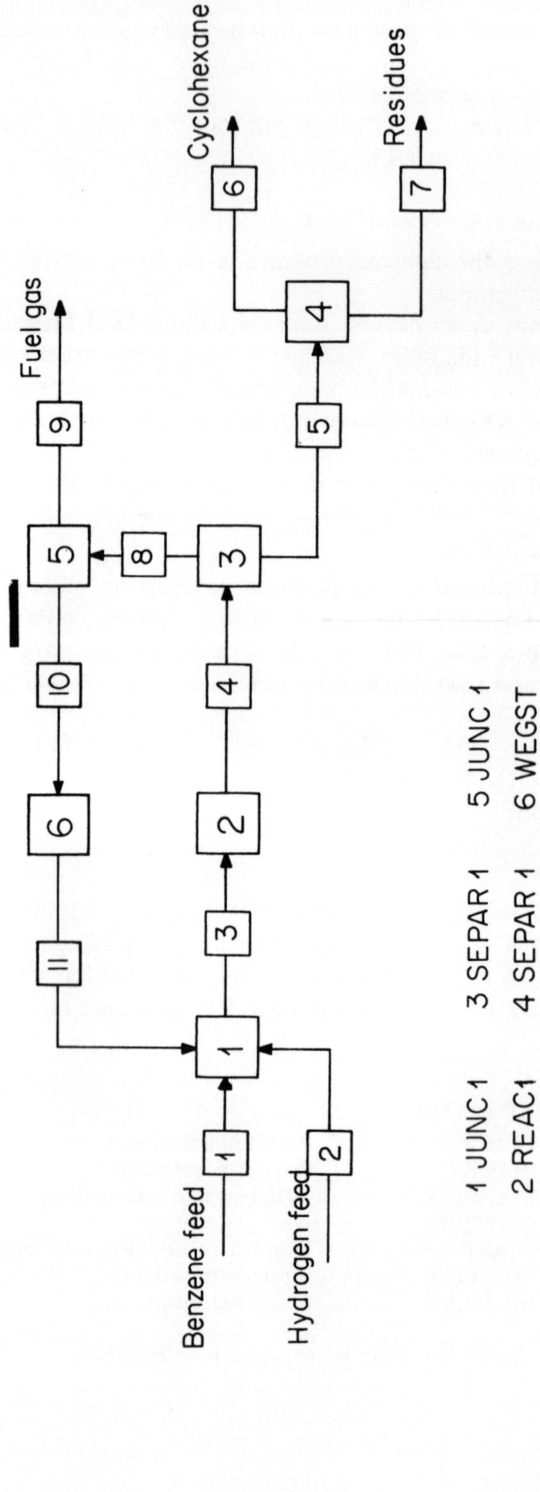

Figure A.1 Cyclohexane information flow.

ALIST Program description

Card	*Notation*
1	Number of runs: NRUNS.
1	Title: TITLE

Card 1 ∇ NRUNS = I, TITLE = '//////////'

Card	*Notation*
1	Existence of recycle, set 0 if none, 1 if exists: NCYCLE,
1	Number of units in calculation order list; NCALC,
1	Number of components: NCOMP.
2	Calculation order of modules is listed: LIST.
3	Total number of streams: NFLOW,
3	Number of initial streams read: IFLOW,
3	Total number of units: NUNIT.
4	KPRINT parameters: KPRINT.
5	Initial stream numbers: NIF,
5	Component identity numbers: IDCP.
6	Stream identity numbers: IDFL.
7	Cost or value/working unit: COST.
8	Molecular weight: CPMOL.

Card 1	∇NCYCLE = I, NCALC = I, NCOMP = I
Card 2	∇LIST = I, I, I, I. . . .
Card 3	∇NFLOW = I, IFLOW = I, NUNIT = I
Card 4	∇KPRINT = I, I, I, I.
Card 5	∇NIF = I, I. . . . , IDCP = I, I, ...
Card 6	∇IDFL = I, I, I. . . .
Card 7	∇COST = X, X, X, X. . . .
Card 8	∇CPMOL = X, X, X, X. . . .

FLOWLIST Initial stream information

Card	*Item*	*Notation*
1	1	Stream number: FLOWI = I,
1	2	Molar flow rate,
1	3	Stream name code number,
1	4	Cost per unit,
1	5	Temperature,
1	6	Pressure,
1	7	Flow of component 1,
1	8	Flow of component 2,
	⋮	
1	JJ	Flow of component NCOMP.

Card 1 ∇FLOWI = I,X,I,X,X,X,X,X,. . .X

Repeat instructions for each independent stream.

UNITLIST Unit description

Card	*Item*	*Notation*
1	1	Unit number: UNITI = I,
1	2	Module type: NTYPE,
1	3	Number of input streams: NIN,
1	4	Number of output streams: NOUT,
1	5	
	⋮	Input stream numbers,
1	4+NIN	
	⋮	Output stream numbers,
1	5+NIN+NOUT	
	⋮	Zeros,
1	20	
1	21	
	⋮	Set by module.
1	30	
Card 1		∇UNITI = I,I,I,I,I. . .I,I. . .I,I*O,X,X. .

Repeat instructions for each module.
See comment cards for data vector format for each unit.

DLIST Number of alterations to initial data

Card	*Notation*	
1	Modification required	MOD
1	Number of unit modifications	NMUN
1	Number of flow modifications	NMFL
Card 1	∇MOD = I, NMUN = I, NMFL = I	

MFLLIST Stream alterations

Card	*Notation*	
1	Flow parameter numbers	IFL
2	Stream numbers	JFL
3	Modifications	VFL

Card 1 ∇IFL = I,I,I,. . .
Card 2 ∇JFL = I,I,I,. . .
Card 3 ∇VFL = X,X,X,. . .

MUNLIST Unit alterations

Card	*Notation*	
1	Unit parameter numbers	IUN
2	Unit numbers	JUN
3	Modifications	VUN

Card 1 ∇IUN = I,I,I. . .
Card 2 ∇JUN = I,I,I. . .
Card 3 ∇VUN = X,X,X. . .

```
      MASTER EXECUTIVE
C     THIS SIMULATOR IS A MODIFIED VERSION OF THE GEMACS
C     SIMULATOR DEVELOPED BY PROF. A. I. JOHNSON AT
C     MCMASTER UNIVERSITY THIS VERSION HAS BEEN
C     PREPARED BY BURGESS, ROBSON AND WELLS
      INTEGER TITLE(18)
      COMMON/UNITS/UNIT(30,50)
      COMMON/FLOWS/FLOW(14,100)
      COMMON/PARAMS/LIST(50),JJ,LOOP,NIN,NOUT,NCALC,
      NFLOW,NUNIT,NCOMP,NU,KPRINT(4),MLOOP,NF(16),IDCP(8),
     1IDFL(100),COST(100),CPMOL(8),NL
      NAMELIST/ALIST/NRUNS,TITLE
      READ(2,ALIST)
      WRITE(3,107)TITLE
      MRUN=1
C     DETERMINES NUMBER OF CASES STUDIED
   10 IF(NRUNS-MRUN)999,20,20
   20 WRITE(3,101)MRUN
      IF(MRUN-1),30,30,40
C     NOW LOAD DATA
   30 CALL DLOAD
      GO TO 50
C     NOW LOAD DATA FOR MODIFIED CASES
   40 CALL DALTER

C     INITIALIZE MODULE COUNTER
C
   50 NL=1
      KTEST=0
      IF(LOOP-999)70,60,70
   60 WRITE(3,102)
      KTEST=1
   70 CONTINUE
   80 WRITE(3,103)LOOP
C     SET UNIT PARAMETERS FOR CURRENT MODULE
   90 NU=LIST(NL)
      NIN=UNIT(3,NU)+0.001
      NOUT=UNIT(4,NU)+0.001
      NTYPE=ABS(UNIT(2,NU))+0.001
C     NOW CALL MODULE
      DO 100 K=1,16
  100 NF(K)=UNIT(4+K,NU)
      CALL MODULE (NTYPE)

C     RESET MODULE COUNTER
  110 NL=NL+1
      IF(NL-NCALC)90,90,120
  120 CONTINUE
      IF(KPRINT(4))160,180,160
  160 DO 170 I=1,NFLOW
      FLOW(1,I)=I
```

Table A.2 Listing of PRIMER.

```
      WRITE(3,104)I
  170 WRITE(3,105)(FLOW(J,I),J=1,JJ)
  180 IF(LOOP-999)130,140,140
  130 NL=1
      GO TO 80
  140 IF(KTEST)150,150,190
  150 WRITE(3,102)
      NL=0
      KTEST=1
      GO TO 110
C     OUTPUT FINAL STREAM ANALYSIS
  190 CALL WRITE(TITLE,MRUN)
      WRITE(3,106)
      MRUN=MRUN-1
      GO TO 10
  101 FORMAT(1H0,10X,' RUN ', IX,15)
  102 FORMAT(1H0,' FINAL FLOW STREAM ANALYSIS')
  103 FORMAT(1H0,' LOOP',15)
  104 FORMAT(1H0,' STREAM',15)
  105 FORMAT(1H ,6F12.4)
  106 FORMAT(1H0,' END OF RUN')
  107 FORMAT(1H1,18A4)
  999 CONTINUE
      CALL EXIT
      END
```

```
C
      SUBROUTINE DLOAD
C     LOADS DATA
C
C     UNIT ARRRY  UNIT(N,I)  I IS THE MODULE NUMBER
C     N=1 MODULE NUMBER
C     N=2 MODULE TYPE
C     N=3 NUMBER OF IMPUT STREAMS TO MODULE NIN
C     N=4 NUMBER OF OUTPUT STREAMS TO MODULE NOUT
C     N=5-20 INPUT AND OUTPUT STREAM NUMBERS
C     N=21-30 EQUIPMENT PARAMETERS FOR A SPECIFIC
C     MODULE
C
C     FLOW ARRAY  FLOW(J,I)  I IS THE STREAM NUMBER
C     J=1 STREAM NUMBER
C     J=2 MOLAR FLOWRATE
C     J=3 STREAM NAME CODE NUMBER
C     J=4 COST / UNIT
C     J=5 TEMPERATURE
C     J=6 PRESSURE
C     J=7 FLOW OF COMPONENT 1
```

Table A.2 Listing of PRIMER—*continued.*

```
C       J=8 FLOW OF COMPONENT 2
C       ETC.
C       J=JJ FLOW OF COMPONENT NCOMP
C
C
C       NCYCLE=0 FOR NO RECYCLE,  =1 FOR RECYCLE
C       NCALC= THE TOTAL NUMBER OF UNITS IN LIST
C       NCOMP= THE NUMBER OF COMPONENTS
C       LIST= CALCULATION ORDER OF MODULES — MAXIMUM
C       OF 50
C       NFLOW= TOTAL NUMBER OF STREAMS
C       IFLOW= NUMBER OF INITIAL STREAMS READ
C       NUNIT= TOTAL NUMBER OF UNITS
C       KPRINT(1)=1 CAUSING THE PRINTING OF  NCOMP
C       NCALC, LIST
C       KPRINT(2)=1 CAUSING THE PRINTING OF        , INITIAL
C       FLOWS
C       KPRINT(3)=1 CAUSING THE PRINTING OF MODULE SETS
C       KPRINT(4)=1 CAUSES THE PRINTING OF INTERMEDIATE
C       FLOW INFORMATION
C       NIF= INITIAL STREAM NUMBERS
C       IDCP=COMPONENT IDENTITY NUMBER
C       IDFL=STREAM IDENTITY NUMBERS
C       COST=COST/UNIT FOR EACH STREAM
C       CPMOL=COMPONENT MOLECULAR WEIGHT
C
C       SET DUMMY ARRAYS TO FACILITATE NAMELIST INPUT
C       AS DIMENSIONED FLOW PERMITS 100 STREAMS
C       AS DIMENSIONED UNIT PERMITS 50 MODULES
        DIMENSION UNIT(30,50),FLOW(14,100),NIF(20)
        COMMON/UNITS/UNIT1(30),UNIT2(30),UNIT3(30),UNIT4(30),
       1UNIT5(30),UNIT6(30),UNIT7(30),UNIT8(30),UNIT9(30),UNIT10(30),
       2UNIT11(30),UNIT12(30),UNIT13(30),UNIT14(30),UNIT15(30),UNIT16
       3(30),
       4UNIT17(30),UNIT18(30),UNIT19(30),UNIT20(30),UNIT21(30),
       5UNIT22(30),UNIT23(30),UNIT24(30),UNIT25(30),UNIT26(30)
       6UNIT27(30),UNIT28(30),UNIT29(30),UNIT30(30),UNIT31(30),
       8UNIT32(30),UNIT33(30),UNIT34(30),UNIT35(30),UNIT36(30),
       8UNIT37(30),UNIT38(30),UNIT39(30),UNIT40(30),UNIT41(30),
       9UNIT42(30),UNIT43(30),UNIT44(30),UNIT45(30),UNIT46(30),
       AUNIT47(30),UNIT48(30),UNIT49(30),UNIT50(30)
        COMMON/FLOWS/FLOW1(14),FLOW2(14),FLOW3(14),FLOW4(14),
       1FLOW5(14),FLOW6(14),FLOW7(14),FLOW8(14),FLOW9(14),
       2FLOW10(14),FLOW11(14),FLOW12(14),FLOW13(14),FLOW14(14),
       3FLOW15(14),FLOW16(14),FLOW17(14),FLOW18(14),FLOW19(14),
       4FLOW20(14),FLOW21(14),FLOW22(14),FLOW23(14),FLOW24(14),
       5FLOW25(14),FLOW26(14),FLOW27(14),FLOW28(14),FLOW29(14),
       6FLOW30(14),FLOW31(14),FLOW32(14),FLOW33(14),FLOW34(14),
       7FLOW35(14),FLOW36(14),FLOW37(14),FLOW38(14),FLOW39(14),
       9FLOW40(14),FLOW41(14),FLOW42(14),FLOW43(14),FLOW44(14),
       AFLOW45(14),FLOW46(14),FLOW47(14),FLOW48(14),FLOW49(14),
       BFLOW50(14),FLOW51(14),FLOW52(14),FLOW53(14),FLOW54(14),
```

Table A.2 Listing of PRIMER—*continued.*

```
     CFLOW55(14),FLOW56(14),FLOW57(14),FLOW58(14),FLOW59(14),
     DFLOW60(14),FLOW61(14),FLOW62(14),FLOW63(14),FLOW64(14),
     EFLOW65(14),FLOW66(14),FLOW67(14),FLOW68(14),FLOW69(14),
     FFLOW70(14),FLOW71(14),FLOW72(14),FLOW73(14),FLOW74(14),
     GFLOW75(14),FLOW76(14),FLOW77(14),FLOW78(14),FLOW79(14),
     HFLOW80(14),FLOW81(14),FLOW82(14),FLOW83(14),FLOW84(14),
     1FLOW85(14),FLOW86(14),FLOW87(14),FLOW88(14),FLOW89(14),
     2FLOW90(14),FLOW91(14),FLOW92(14),FLOW93(14),FLOW94(14),
     3FLOW95(14),FLOW96(14),FLOW97(14),FLOW98(14),FLOW99(14),
     4FLOW100(14)
      COMMON/PARAMS/LIST(50),JJ,LOOP,NIN,NOUT,NCALC,
     1NFLOW,NUNIT,NCOMP,NU,KPRINT(4),MLOOP,NF(16),IDCP(8),
     2IDFL(100),COST(100),CPMOL(8)
      NAMELIST/BLIST/NCYCLE,NCALC,NCOMP,LIST,NFLOW,
     1IFLOW,NUNIT,KPRINT,NIF,IDCP,IDFL,COST,CPMOL
      NAMELIST/UNITLIST/UNIT1,UNIT2,UNIT3,UNIT4,UNIT5,UNIT6,
     1UNIT7,UNIT8,UNIT9,UNIT10,UNIT11,UNIT12,UNIT13,UNIT14,
     2UNIT15,UNIT16,UNIT17,UNIT18,UNIT19,UNIT20,UNIT21,UNIT22,
     3UNIT23,UNIT24,UNIT25,UNIT26,UNIT27,UNIT28,UNIT29,UNIT30,
     4UNIT31,UNIT32,UNIT33,UNIT34,UNIT35,UNIT36,UNIT37,UNIT38,
     5UNIT39,UNIT40,UNIT41,UNIT42,UNIT43,UNIT44,UNIT45,UNIT46,
     6UNIT47,UNIT48,UNIT49,UNIT50
      NAMELIST/FLOWLIST/FLOW1,FLOW2,FLOW3,FLOW4,FLOW5,
     1FLOW6,FLOW7,FLOW8,FLOW9,FLOW10,FLOW11,FLOW12,
     2FLOW13,FLOW14,FLOW15,FLOW16,FLOW17,FLOW18,FLOW19,
     3FLOW20,FLOW21,FLOW22,FLOW23,FLOW24,FLOW25,FLOW26,
     4FLOW27,FLOW28,FLOW29,FLOW30,FLOW31,FLOW32,FLOW33,
     4FLOW34,FLOW35,FLOW36,FLOW37,FLOW38,FLOW39,FLOW40,
     5FLOW41,FLOW42,FLOW43,FLOW44,FLOW45,FLOW46,FLOW47,
     6FLOW48,FLOW49,FLOW50,FLOW51,FLOW52,FLOW53,FLOW54,
     7FLOW55,FLOW56,FLOW57,FLOW58,FLOW59,FLOW60,FLOW61,
     8FLOW62,FLOW63,FLOW64,FLOW65,FLOW66,FLOW67,FLOW68,
     9FLOW69,FLOW70,FLOW71,FLOW72,FLOW73,FLOW74,FLOW75,
     AFLOW76,FLOW77,FLOW78,FLOW79,FLOW80,FLOW81,FLOW82,
     BFLOW83,FLOW84,FLOW85,FLOW86,FLOW87,FLOW88,FLOW89,
     CFLOW90,FLOW91,FLOW92,FLOW93,FLOW94,FLOW95,FLOW96,
     DFLOW97,FLOW98,FLOW99,FLOW100
      EQUIVALENCE (UNIT,UNIT1),(FLOW,FLOW1)
C     INITIALISE ARRAYS UNIT AND FLOW
      DO 10 I=1,30
      DO 10 J=1,50
   10 UNIT(I,J)=0.
      DO 20 I=1,14
      DO 20 J=1,100
   20 FLOW(I,J)=0.
      READ(2,BLIST)
C
C     READ NO. OF UNITS IN CALCULATION ORDER AND NO
C     OF COMPONENTS
C
      IF(KPRINT(1))30,40,30
```

Table A.2 Listing of PRIMER—*continued.*

```
   30 WRITE(3,102)NCALC,NCOMP
C     IS RECYCLE REQUIRED?
   40 IF(NCYCLE)50,50,60
   50 LOOP=999
      MLOOP=0
      GO TO 61
   60 LOOP=1
      MLOOP=1
C
C     READ CALCULATION ORDER
C
   61 IF(KPRINT(1))70,80,70
   70 WRITE(3,103)(LIST(I),I=1,NCALC)
C
C     READ INITIAL STREAMS AND SET LENGTH OF FLOW
C     VECTOR=JJ
C
   80 JJ=NCOMP+6
      IF(KPRINT(2))90,100,90
   90 WRITE(3,104)IFLOW
  100 READ(2,FLOWLIST)
      DO 81 I=1,NFLOW
      FLOW(3,I)=IDFL(I)
      FLOW(4,I)=COST(I)
   81 CONTINUE
      DO 120 I=1,IFLOW
      IF(KPRINT(2))110,120,110
  110 WRITE(3,105)NIF(I)
      WRITE(3,106)(FLOW(J,NIF(I)),J=1,JJ)
  120 CONTINUE
C
C     READING IN UNIT SETS ,NUNIT IN NUMBER
C
      IF(KPRINT(3))130,140,130
  130 WRITE(3,107)NUNIT
      READ(2,UNITLIST)
  140 DO 160 I=1, NUNIT
      MM=UNIT(1,I)
      IF(KPRINT(3))150,160,150
  150 WRITE(3,108)MM
      WRITE(3,109)(UNIT(N,I),N=1,30)
  160 CONTINUE
  102 FORMAT(1HO,' NO. OF UNITS IN CALCULATION ORDER
     1LIST =',I5/' NO. OF COMPONENTS =',I5/' CALCULATION
     1ORDER LIST IS')
  103 FORMAT(1X,5I5,5X,5I5)
  104 FORMAT(1H0,' NO. OF INITIAL FLOWS READ =',I5)
  105 FORMAT(1H0,'  STREAM', I5)
  106 FORMAT(1X,6F12.4)
  107 FORMAT(1H0,' NO. OF UNIT SETS READ =',I5)
  108 FORMAT(1H0,' UNIT',I5)
```

Table A.2 Listing of PRIMER—*continued.*

```
  109 FORMAT(1X,4F12.4)
      RETURN
      END
      SUBROUTINE WRITE(TITLE,MRUN)
      INTEGER FLNAME(100),CPNAME(80),TITLE(18)
      DIMENSION NK(3),TMASS(3),AVMOL(3),CPMASS(3)
      COMMON/UNITS/UNIT(30,50)
      COMMON/FLOWS/FLOW(14,100)
      COMMON/PARAMS/LIST(50),JJ,LOOP,NIN,NOUT,NCALC,
     1NFLOW,NUNIT,NCOMP,NU,KPRINT(4),MLOOP,NF(16),IDCP(8),
     1IDFL(100),COST(100),CPMOL(8)
C     COMPONENT NAME IDENTITY NUMBERS
C     1. HYDROGEN          2. METHANE          3. ETHANE
C     4. PROPANE           5. ETHYLENE         6. PROPYLENE
C     7. CYCLOPENTANE      8. CYCLOHEXANE      9. BENZENE
C    10. TOLUENE          11. O-XYLENE        12. M-XYLENE
C    13. P-XYLENE  14. ETHYLBENZENE  15. HYDRODEN SULPHIDE
C    16. SULPHUR DIOXIDE  17. WATER           18. AMMONIA
C    19. ACETYLENE        20. N-HEXANE
      DATA CPNAME(1)/'HYDROGEN                          METHANE
     1ETHANE                               PROPANE
     2ETHYLENE                            PROPYLENE
     3CYCLOPENTANE                      CYCLOHEXANE
     4BENZENE                          TOLUENE
     5O-XYLENE                       M-XYLENE
     6P-XYLENE                      ETHYLBENZENE
     7HYDROGEN SULFIDE SULPHUR     DIOXIDE WATER
     8AMMONIA                     ACETYLENE
     9N-HEXANE                      '/
                                  RAW MATERIAL   PRODUCT
     1
     2
     3
     4
     5
     6                     '/
      WRITE(3,201)TITLE,MRUN
      DO 200 II=1,NFLOW,3
      DO 220 I=1,3
  220 NK(I)=(FLOW(3,II-1+I)-1)*4
      DO 221 LL=1,3
  221 TMASS(LL)=0.
      III=II+2
      IF(III.GT.NFLOW)GO TO 230
      GO TO 240
  230 III=NFLOW
  240 WRITE(3,202)(FLOW(1,I),I=II,III)
      KJ=III-II+1
      WRITE(3,203)(FLNAME(NK(J)+1),FLNAME(NK(J)+2),FLNAME
     1(NK(J)+3),FLNAME(NK(J)+4),J=1,KJ)
      WRITE(3,209)
```

Table A.2 Listing of PRIMER—*continued.*

```
      DO 210 J=1,NCOMP
      JJJ=J+6
      DO 222 I=II,III
      L=I-II+1
      CPMASS(L)=FLOW(JJJ,I)*CPMOL(J)
  222 TMASS(L)=TMASS(L)+FLOW(JJJ,I)*CPMOL(J)
      NJ=(IDCP(J)-1)*4+1
      WRITE(3,207)J,(CPNAME(NI),NI=NJ,NJ+3),CPMOL(J),(FLOW
     1(JJJ,I),CPMASS(I-II+1),I=II,III)
  210 CONTINUE
      WRITE(3,208)(FLOW(2,I),TMASS(I-II+1),I=II,III)
      DO 223 I=1,KJ
  223 AVMOL(I)=TMASS(I)/FLOW(2,I+II-1)
      WRITE(3,211)(AVMOL(I),I=1,KJ)
      WRITE(3,204)(FLOW(4,I),I=II,III)
      WRITE(3,205)(FLOW(5,I),I=II,III)
      WRITE(3,206)(FLOW(6,I),I=II,III)
      WRITE(3,212)
  200 CONTINUE
  201 FORMAT(1H1,T20,18A4,T50,'CASE ',I5)
  202 FORMAT(1H0,'STREAM NUMBER',T25,3(14X,F3.0,13X))
  203 FORMAT(1H ,'STREAM NAME',T25,3(7X,4A4,7X))
  204 FORMAT(1H ,'COST/UNIT',T25,3(12X,F6.2,12X))
  205 FORMAT(1H ,'TEMPERATURE',T25,3(12X,F6.2,12X))
  206 FORMAT(1H ,'PRESSURE',T25,3(12X,F6.2,12X))
  207 FORMAT(1H ,I1,T4,4A4,F3.0,T25,3(3X,F10.4,2X,F12.2,3X))
  208 FORMAT(1H0,'TOTAL FLOW',T25,3(3X,F10.4,2X,F12.2,3X))
  209 FORMAT(1H0,T2,'COMPONENT',T20,'M.W.',T25,3(7X,'MOLES',
     17X,'MASS',7X))
  211 FORMAT(1H0,'AVERAGE MOL.WT.',T25,3(12X,F6.1,12X))
  212 FORMAT(1H0,///,'**********************************************
     1*********************',///)
      RETURN
      END



      SUBROUTINE DALTER
C
C     LOADS MODIFIED DATA
C     MODIFICATIONS TO
C         STREAMS AND MODULES IF MOD=0
C         STREAMS ONLY IF MOD=1
C         MODULES ONLY IF MOD=2
C     100 FLOWS AND 100 UNIT MODIFICATIONS ARE ALLOWED
C     IUN= MODULES ELEMENTS TO BE MODIFIED
C     JUN= NUMBERS OF MODIFIED MODULES
C     VUN= MODIFIED VALUE OF MODULE ELEMENT
```

Table A.2 Litsing of PRIMER—*continued.*

```
C       IFL=FLOW ELEMENTS TO BE MODIFIED
C       JFL= NUMBERS OF MODIFIED FLOWS
C       VFL= MODIFIED VALES OF STREAM ELEMENTS
C
        DIMENSION IFL(100),JFL(100),VFL(100),IUN(100),JUN(100),
       1VUN(100)
        COMMON/UNITS/UNIT(30,50)
        COMMON/FLOWS/FLOW(14,100)
        COMMON/PARAMS/LIST(50),JJ,LOOP,NIN,NOUT,NCALC,
       1NFLOW,NUNIT,NCOMP,NU,KPRINT(4),MLOOP,NF(16),IDCP(8),
       1IDFL(100),COST(100),CPMOL(8)
        NAMELIST/DLIST/MOD,NMUN,NMFL
        NAMELIST/MFLLIST/IFL,JFL,VFL
        NAMELIST/MUNLIST/IUN,JUN,VUN
C
        DO 10 K=1,100
        IFL(K)=0
        JFL(K)=0
        VFL(K)=0
        IUN(K)=0
        JUN(K)=0
        VUN(K)=0
     10 CONTINUE
        IF(MLOOP)40,40,20
     20 LOOP=1
     40 READ(2,DLIST)
        IF(MOD.EQ.1)GO TO 60
C       INITIALIZE MODIFIED DATA ARRAYS FOR UNITS AND
C       FLOWS
        READ(2,MUNLIST)
        DO 51 I=1,NMUN
     51 UNIT(IUN(I),JUN(I))=VUN(I)
        IF(MOD.EQ.2)GO TO 70
     60 READ(2,MFLLIST)
        DO 61 J=1,NMFL
     61 FLOW(IFL(J),JFL(J))=VFL(J)
     70 CONTINUE
        IF(NMFL.GT.100.OR.NMUN.GT.100)GO TO 900
        IF(NMFL.EQ.0)GO TO 600
C       FLOW MODIFICATIONS
        WRITE(3,101)NMFL
        WRITE(3,102)(JFL(J),IFL(J),VFL(J),J=1,NMFL)
        DO 500 J=1,NMFL
        IF(JFL(J+1).EQ.JFL(J))GO TO 500
        WRITE(3,103)JFL(J)
        WRITE(3,104)(FLOW(I,JFL(J)),I=1,JJ)
    500 CONTINUE
    600 IF(NMUN.EQ.0)GO TO 80
C
C       UNIT MODIFICATIONS
C
```

Table A.2 Listing of PRIMER—*continued.*

```
      WRITE(3,105)NMUN
      WRITE(3,106)(JUN(J),IUN(J),VUN(J),J=1,NMUN)
      DO 700 J=1,NMUN
      IF(JUN(J+1).EQ.JUN(J))GO TO 700
      WRITE(3,107)JUN(J)
      WRITE(3,104)(UNIT(I,JUN(J)),I=1,30)
  700 CONTINUE
      GO TO 80
  900 WRITE(3,108)
   80 CONTINUE
  101 FORMAT('0NUMBER OF FLOW MODIFICATIONS IS ',I5)
  102 FORMAT('0STREAM NUMBER ',I5,' HAS ELEMENT NUMBER
     1',I5,' MODIFIED--NEW VALUE IS ',F12.4)
  103 FORMAT('0NEW DATA BLOCK FOR STREAM NUMBER
     1',I5,' IS')
  104 FORMAT(5F12.4)
  105 FORMAT(1H0,' NUMBER OF UNIT MODIFICATIONS IS ',I5)
  106 FORMAT(1H0,' UNIT NUMBER',I5,' HAS ELEMENT NUMBER
     1',15,' MODIFIED---NEW VALUE IS ',F12.4)
  107 FORMAT(1HO,' NEW DATA BLOCK FOR UNIT NUMBER
     1',I5,' IS')
  108 FORMAT(1H0,' ERROR-----EITHER NO. OF FLOW OR UNIT
     1MODIFICATIONS EXCEEDS MAXIMUM ALLOWED')
      RETURN
      END
```

```
      SUBROUTINE MODULE(NT)
C     IDENTIFIES UNIT TYPE AND CALLS REQUIRED MODULE
      GO TO(1,2,3,4,5,6,7,8,9,10,11,12,13,14,15,16,17,18,19,20),NT
    1 CALL CONST1
      GO TO 50
    2 CALL CSCADE
      GO TO 50
    3 CALL DIFF
      GO TO 50
    4 CALL JUNC1
      GO TO 50
    5 CALL LOOP1
      GO TO 50
    6 CALL REAC1
      GO TO 50
    7 CALL SEPAR1
      GO TO 50
    8 CALL SEPAR2
      GO TO 50
    9 CALL TEST
      GO TO 50
```

Table A.2 Listing of PRIMER—*continued.*

```
   10 CALL WEGST
      GO TO 50
   11 CALL EXTRA1
      GO TO 50
   12 CALL EXTRA2
      GO TO 50
   13 CALL EXTRA3
      GO TO 50
   14 CALL EXTRA4
      GO TO 50
   15 GO TO 50
   16 GO TO 50
   17 GO TO 50
   18 GO TO 50
   19 GO TO 50
   20 GO TO 50
   50 RETURN
      END
```

```
      SUBROUTINE JUNC1
C
C     SIMPLE MIXER/SPLITTER
C     UNIT J=1-20 STANDARD
C          J=21-20+NOUT FRACTION OF TOTAL INPUT IN
C          1ST,2ND--NOUT OUTPUT STREAM
C
      DIMENSION TOTAL(14)
      COMMON/UNITS/UNIT(30,50)
      COMMON/FLOWS/FLOW(14,100)
      COMMON/PARAMS/LIST(50),JJ,LOOP,NIN,NOUT,NCALC,
     1NFLOW,NUNIT,NCOMP,NU,KPRINT(4),MLOOP,NF(16),IDCP(8),
     1IDFL(100),COST(100),CPMOL(8)
      NN=NOUT+NIN
      LL=NIN+1
C     INPUT FLOWS SUMMED
      DO 1 J=7,JJ
      TOTAL(J)=0.
      DO 1 I=1, NIN
    1 TOTAL(J)=TOTAL(J)+FLOW(J,NF(I))
C     OUTPUT FLOWS CALULATED
      DO 2 I=LL,NN
      II=21+I-LL
      FLOW(5,NF(I))=FLOW(5,NF(1))
      FLOW(6,NF(I))=FLOW(6,NF(1))
      FLOW(2,NF(I))=0.
      DO 2 J=7,JJ
```

Table A.2 Listing of PRIMER—*continued.*

```
      FLOW(J,NF(I)) =TOTAL(J)*UNIT(II,NU)
    2 FLOW(2,NF(I)) =FLOW(2,NF(I)) +FLOW(J,NF(I))
      RETURN
      END




      SUBROUTINE LOOP1
C
C     RESET LOOP COUNTER
C
      COMMON/UNITS/UNIT(30,50)
      COMMON/FLOWS/FLOW(14,100)
      COMMON/PARAMS/LIST(50),JJ,LOOP,NIN,NOUT,NCALC,
     1NFLOW,NUNIT,NCOMP,NU,KPRINT(4),MLOOP,NF(16),IDCP(8),
     1IDFL(100),COST(100),CPMOL(8)
      IF(UNIT(21,NU))2,2,1
    1 LOOP=1
      UNIT(21,NU)=-1
      GO TO 3
    2 UNIT(21,NU)=1
    3 RETURN
      END




      SUBROUTINE REAC1
C
C     REACTOR MODULE
C     CONVERSION OF KEY COMPONENT KNOWN
C     STOICHIOMETRIC COEFFICIENTS KNOWN
C     UNIT J=1-20 STANDARD
C            J=21 FRACTIONAL CONVERSION
C            J=22 KEY COMPONENT NUMBER
C            J=23-22+NCOMP STOICHIOMETRIC COEFFICIENTS
C
      COMMON/UNITS/UNIT(30,50)
      COMMON/FLOWS/FLOW(14,100)
      COMMON/PARAMS/LIST(50),JJ,LOOP,NIN,NOUT,NCALC,
     1NFLOW,NUNIT,NCOMP,NU,KPRINT(4),MLOOP,NF(16),IDCP(8),
     1IDFL(100),COST(100),CPMOL(8)
      FLOW(5,NF(2)) =FLOW(5,NF(1))
      FLOW(6,NF(2)) =FLOW(6,NF(1))
      FLOW(2,NF(2)) =0.
      KCOMP=UNIT(22,NU)+6
      DO 1 J=7,JJ
```

Table A.2 Listing of PRIMER—*continued.*

```
      FLOW(J,NF(2))=FLOW(J,NF(1))+FLOW(KCOMP,NF(1))*UNIT
     1(16+J,NU)*UNIT(21,NU)
      FLOW(2,NF(2))=FLOW(2,NF(2))+FLOW(J,NF(2))
    1 CONTINUE
      RETURN
      END
```

```
      SUBROUTINE SEPAR1
C
C     SIMPLE SEPARATOR TYPE 1
C     COMPONENT FRACTIONS OF INPUT FLOW WHICH GO TO
C     FIRST OUTPUT FLOW KNOWN
C     UNIT J=1-20 STANDARD
C     UNIT J=21-20+NCOMP FRACTIONS
C
      COMMON/UNITS/UNIT(30,50)
      COMMON/FLOWS/FLOW(14,100)
      COMMON/PARAMS/LIST(50),JJ,LOOP,NIN,NOUT,NCALC,
     1NFLOW,NUNIT,NCOMP,NU,KPRINT(4),MLOOP,NF(16),IDCP(8),
     1IDFL(100),COST(100),CPMOL(8)
      DO 2 I=2,3
      FLOW(5,NF(I))=FLOW(5,NF(1))
      FLOW(6,NF(I))=FLOW(6,NF(1))
    2 CONTINUE
      FLOW(2,NF(2))=0.
      FLOW(2,NF(3))=0.
      DO 1 J=7,JJ
C     CALCULATE FIRST OUTPUT FLOW
      FLOW(J,NF(2))=FLOW(J,NF(1))*UNIT(J+14,NU)
      FLOW(2,NF(2))=FLOW(2,NF(2))+FLOW(J,NF(2))
C     CALCULATE SECOND OUTPUT FLOW
      FLOW(J,NF(3))=FLOW(J,NF(1))-FLOW(J,NF(2))
      FLOW(2,NF(3))=FLOW(2,NF(3))+FLOW(J,NF(3))
    1 CONTINUE
      RETURN
      END
```

```
      SUBROUTINE WEGST
C
C     CONVERGENCE TEST ROUTINE WITH WEGSTEIN
C     CONVERGENCE PROMOTER
```

Table A.2 Listing of PRIMER—*continued.*

```
C       UNIT J=1-20 STANDARD
C            J=21 MAXIMUM NUMBER OF LOOPS
C            J=22 FRACTIONAL TOLERANCE
CC           J=23 LOOP WHEN WEGSTEIN FIRST USED
C            J=23-22+NCOMP STORE LATEST COMPONENT FLOW
C            VALUES
C
        DIMENSION SAVE(16)
        COMMON/UNITS/UNIT(30,50)
        COMMON/FLOWS/FLOW(14,100)
        COMMON/PARAMS/LIST(50),JJ,LOOP,NIN,NOUT,NCALC,
       1NFLOW,NUNIT,NCOMP,NU,KPRINT(4),MLOOP,NF(16),IDCP(8),
       1IDFL(100),COST(100),CPMOL(8),NL
        JTEST=0
        FLOW(5,NF(2))=FLOW(5,NF(1))
        FLOW(6,NF(2))=FLOW(6,NF(1))
        IF(LOOP-999)1,13,13
      1 WRITE(3,9)LOOP
        IF(LOOP.EQ.1)GO TO 7
        DO 3 J=7,JJ
        IJ=J-6
C       COMPARE FLOW WITH PREVIOUS VALUE
        IF(FLOW(J,NF(1)))41,42,41
     41 DIFF=ABS((FLOW(J,NF(1))-FLOW(J,NF(2)))/FLOW(J,NF(1)))
        GO TO 43
     42 DIFF=ABS(FLOW(J,NF(1))-FLOW(J,NF(2)))
     43 WRITE(3,10)IJ,DIFF
        IF(DIFF-UNIT(22,NU))3,3,2
      2 JTEST=1
      3 CONTINUE
        IF(JTEST)4,4,5
      4 WRITE(3,11)
        LOOP=999
        GO TO 13
      5 WRITE(3,12)
        ML=UNIT(21,NU)+0.001
        IF(LOOP-ML)7,6,6
      6 LOOP=999
        GO TO 13
     13 RETURN
      7 LW=UNIT(23,NU)+0.001
        IF(LOOP-LW)15,14,14
     14 WRITE(3,16)
     15 DO 17 J=7,JJ
        JK=J-6
        KK=NCOMP+JK
        WN=FLOW(J,NF(1))
        WO=SAVE(JK)
        WC1=FLOW(J,NF(2))
        WC2=SAVE(KK)
        IF(LOOP-LW)25,18,18
```

Table A.2 Listing of PRIMER—*continued.*

```
22 IF(IWEG)19,19,20
19 WRITE(3,21)
20 CONTINUE
   SAVE(JK)=WN
   SAVE(KK)=WC1
   FLOW(J,NF(2))=WWN
   JC=J-6
   WRITE(3,23)JC,PHI,Q
   GO TO 17
25 SAVE(JK)=WN
   SAVE(KK)=WC1
   FLOW(J,NF(2))=FLOW(J,NF(1))
17 CONTINUE
   FLOW(2,NF(2))=0.
   DO 27 J=7,JJ
27 FLOW(2,NF(2))=FLOW(2,NF(2))+FLOW(J,NF(2))
   WRITE(3,28)
   WRITE(3,29)(FLOW(J,NF(2)),J=1,JJ)
   LOOP=LOOP+1
   NL=NL-IFIX(UNIT(20,NU))-1
   GO TO 30
18 IWEG=1
   IF(WN-WO)31,39,31
31 IF(WC1-WC2)32,39,32
32 PHI=(WC1-WC2)/(WN-WO)
   IF(PHI-1.)33,39,33
33 Q=1./(1.-PHI)
   IF(Q-10.)35,35,34
34 Q=10.
   GO TO 37
35 IF(Q+10.)36,37,37
36 CONTINUE
37 WWN=Q*WC1+(1.-Q)*WN
   GO TO 40
39 WWN=WN
   Q=0.
   PHI=0.
   IWEG=-1
40 GO TO 22
 9 FORMAT(1H0,' END OF LOOP',I5)
10 FORMAT(1H0,'FRACTIONAL CHANGE IN COMPONENT',I5,
  1'IS',3X,F10.8)
11 FORMAT(1H0,' ***CONVERGED***')
12 FORMAT(1H0,' ***NOT CONVERGED***')
16 FORMAT(1H0,T6,'J',T16,' PHI',T30,' Q')
21 FORMAT(1H0,' WEGSTEIN FAILS')
23 FORMAT(1H ,T6,I1,T10,F12.8,T24,F12.8)
28 FORMAT(1H0,' NEW VALUE OF FLOW')
29 FORMAT(1H ,6F12.4)
30 RETURN
   END
```

Table A.2 Listing of PRIMER—*continued.*

```
&ALIST
NRUNS=1, TITLE= 'CYCLOHEXANE  PLANT'
&END
&BLIST
NCYCLE=1, NCALC=6, NCOMP=5
LIST=1, 2, 3, 5, 6, 4
NFLOW=11, 1FLOW=2, NUNIT=6
KPRINT=1, 1, 1
NIF=1, 2, 1DCP=1, 3, 8, 9, 10
IDFL=2, 2, 1, 1, 1, 3, 3, 1, 3, 1, 1
COST=0
CPMOL=2., 16., 84., 78., 92.
&END
&FLOWLIST
FLOW1=1, 100., 0, 0, 0, 0, 0, 0, 0, 99.5, 0.5
FLOW2=2, 500., 0, 0, 0, 0, 475., 25.
&END
&UNITLIST
UNIT1=1, 4, 3, 1, 1, 2, 11, 3, 12*0, 1.
UNIT2=2, 6, 1, 1, 3, 4, 14*0, 0.999, 3, -3, 0, 1, -1
UNIT3=3, 7, 1, 2, 4, 8, 5, 13*0, 1., 1.
UNIT4=4, 7, 1, 2, 5, 6, 7, 13*0, 0, 0, 0.999, 0.001
UNIT5=5, 4, 1, 2, 8, 9. 10, 13*0, 0.3
UNIT6=6, 10, 1, 1, 10, 11, 14*0, 50, 0.001, 4
&END
```

Table A.3 Cyclohexane plant.

STREAM NUMBER		1.		2.		3.	
STREAM NAME		RAW MATERIAL		RAW MATERIAL			
COMPONENT	MOL.WT.	MOLES	MASS	MOLES	MASS	MOLES	MASS
1 HYDROGEN	2.	0.0000	0.00	475.0000	950.00	887.5298	1775.06
2 METHANE	16.	0.0000	0.00	25.0000	400.00	83.3333	1333.33
3 CYCLOHEXANE	84.	0.0000	0.00	0.0000	0.00	0.0000	0.00
4 BENZENE	78.	99.5000	7761.00	0.0000	0.00	99.5000	7761.00
5 TOLUENE	92.	0.5000	46.00	0.0000	0.00	0.5000	46.00
TOTAL FLOW		100.0000	7807.00	500.0000	1350.00	1070.8632	10915.39
AVERAGE MOL.WT.		78.1		2.7		10.2	
COST/UNIT		0.00		0.00		0.00	
TEMPERATURE		0.00		0.00		0.00	
PRESSURE		0.00		0.00		0.00	

****

STREAM NUMBER		4.		5.		6.	
STREAM NAME						PRODUCT	
COMPONENT	MOL.WT.	MOLES	MASS	MOLES	MASS	MOLES	MASS
1 HYDROGEN	2.	589.3283	1178.66	0.0000	0.00	0.0000	0.00
2 METHANE	16.	83.3333	1333.33	0.0000	0.00	0.0000	0.00
3 CYCLOHEXANE	84.	99.4005	8349.64	99.4005	8349.64	99.3011	8341.29
4 BENZENE	78.	0.0995	7.76	0.0995	7.76	0.0001	0.01
5 TOLUENE	92.	0.5000	46.00	0.5000	46.00	0.0000	0.00
TOTAL FLOW		772.6617	10915.39	100.0000	8403.40	99.3012	8341.30
AVERAGE MOL.WT.		14.1		84.0		84.0	
COST/UNIT		0.00		0.00		0.00	
TEMPERATURE		0.00		0.00		0.00	
PRESSURE		0.00		0.00		0.00	

****

Table A.4 Cyclohexane plant.

STREAM NUMBER		7.		8.		9.	
STREAM NAME		PRODUCT				PRODUCT	
COMPONENT	MOL.WT.	MOLES	MASS	MOLES	MASS	MOLES	MASS
1 HYDROGEN	2.	0.0000	0.00	589.3283	1178.66	176.7985	353.60
2 METHANE	16.	0.0000	0.00	83.3333	1333.33	25.0000	400.00
3 CYCLOHEXANE	84.	0.0994	8.35	0.0000	0.00	0.0000	0.00
4 BENZENE	78.	0.0994	7.75	0.0000	0.00	0.0000	0.00
5 TOLUENE	92.	0.5000	46.00	0.0000	0.00	0.0000	0.00
TOTAL FLOW		0.6988	62.10	672.6617	2511.99	201.7985	753.60
AVERAGE MOL.WT.		88.9		3.7		3.7	
COST/UNIT		0.00		0.00		0.00	
TEMPERATURE		0.00		0.00		0.00	
PRESSURE		0.00		0.00		0.00	

****

STREAM NUMBER		10.		11.			
STREAM NAME							
COMPONENT	MOL.WT.	MOLES	MASS	MOLES	MASS	MOLES	MASS
1 HYDROGEN	2.	412.5298	825.06	412.5298	825.06		
2 METHANE	16.	58.3333	933.33	58.3333	933.33		
3 CYCLOHEXANE	84.	0.0000	0.00	0.0000	0.00		
4 BENZENE	78.	0.0000	0.00	0.0000	0.00		
5 TOLUENE	92.	0.0000	0.00	0.0000	0.00		
TOTAL FLOW		470.8632	1758.39	470.8632	1758.39		
AVERAGE MOL.WT.		3.7		3.7			
COST/UNIT		0.00		0.00			
TEMPERATURE		0.00		0.00			
PRESSURE		0.00		0.00			

****

Table A.4 Cyclohexane plant—*continued.*

Index